ALSO AVAILABLE BY BERND HEINRICH

Bumblebee Economics
One Man's Owl
A Year in the Maine Woods
The Trees in My Forest
Mind of the Raven
Why We Run
Winter World
The Snoring Bird
Summer World
The Nesting Season
Life Everlasting
The Homing Instinct

RAVENS

IN

WINTER

BERND HEINRICH

ILLUSTRATIONS BY THE AUTHOR

SIMON & SCHUSTER PAPERBACKS
NEW YORK LONDON TORONTO SYDNEY NEW DELHI

Simon & Schuster Paperbacks
A Division of Simon & Schuster, Inc.
1230 Avenue of the Americas
New York, NY 10020

First Simon & Schuster trade paperback edition October 2014

SIMON & SCHUSTER PAPERBACKS and colophon
are registered trademarks of Simon & Schuster, Inc.

For information about special discounts for bulk purchases,
please contact Simon & Schuster Special Sales at 1-866-506-1949
or business@simonandschuster.com.

The Simon & Schuster Speakers Bureau can bring authors
to your live event. For more information or to book an event,
contact the Simon & Schuster Speakers Bureau at 1-866-248-3049
or visit our website at www.simonspeakers.com.

Cover illustration by Bernd Heinrich

Manufactured in the United States of America

10 9 8 7 6 5 4 3 2 1

Library of Congress Cataloging-in-Publication Data

ISBN 978-1-4767-9456-3 (pbk)
ISBN 978-1-4767-9457-0 (ebook)

To all the raven mainiacs
who answered the call

CONTENTS

Illustrations following page 128

NOTE ON THE NEW EDITION

Ravens in Winter is now celebrating its twenty-fifth anniversary. It is a scientific detective story derived from a commonplace sighting I made on October 18, 1984, in the Maine woods. I had observed the puzzling behavior of a large group of ravens that I thought might have been sharing a prized food bonanza—a moose carcass. But such sharing made absolutely no sense to me. It went against the grain of everything I had learned in my pursuit of classical biology. I was then officially an insect biologist with fourteen years of studying the physiology and behavior of bumblebees just recently behind me. Biology, the study of life, is all about finding generalities behind often seemingly idiosyncratic differences: when I saw those otherwise highly aggressive and territorial birds all sharing the same food bonanza, I could not help but think they held some profound and interesting secret, maybe even one that could apply to humans.

I had always been interested in birds, but felt we knew them well. Yet my observation that day seemed totally at odds with everything I knew. There were several possibilities, but there had to be one answer and that answer would reveal something new and wonderful about ravens and perhaps about animals in general. The question was so important that it would need to be solved and I wanted to leave a record of how it was solved. I thought, however, that the journey might be even more interesting than the destination. It would definitely be difficult because in the Maine woods ravens were not only uncommon but absurdly alert and shy of humans. I felt that the process of solving a scientific puzzle, using techniques that were available to almost everyone—because I had no special resources myself—would make for interesting reading and show how science is done or can be done by anyone venturing to tread on new ground. My task was to sort out and reject various hypotheses by experimentation and observation, as is the process of science.

The challenges were quite daunting, but the prize lured me on, and not always in the right direction. The false starts were ultimately essential, because each one narrowed the possibilities, leading me, hopefully, in the right direction. Not knowing anything about ravens proved to be a blessing because I was not biased by previous knowledge but informed by it. My writing was directly done from my research in the field and because I was typing-illiterate I did not use a computer. Yet when I submitted a draft of my notes for the book to the late Anne Freedgood, my editor at Summit Books, she was excited and worked on it over the weekend, handing it back to me all marked up in pencil. I thank her still for her encouragement and guidance.

Little did I suspect what a Pandora's box my book would open. In the thirty years since its publication there has been an explosion of research on ravens and other birds of the crow family, and the findings are nothing if not astounding. I feel a new window has been opened into the minds of animals, one never before suspected.

I was highly honored when the publisher agreed to reprint *Ravens in Winter* after its long "hibernation" in obscurity overshadowed by more recent knowledge-based books, including my own *Mind of the Raven*. Although the latter became a winner of the John Burroughs Medal for natural history writing, to me the more simple *Ravens in Winter* is by far the more important book, because it represents the passions, joys, and sometimes heartache of what seemed to me at the time a life poured wholeheartedly into an endeavor equivalent to Sir Edmund Hillary tackling Mount Everest for the first time. I had not done anything like it before, or since.

When considering the reissue of this book, the question came up of what to include: whether more, or less, of anything. I thought of perhaps deleting the appendix, which includes my hard-won data. I felt that this data in charts and graphs might have been off-putting to many readers because it gave a technical "flavor" to the book, which is not its thrust. But I decided to keep it in, to preserve the original.

When I originally wrote the book, I needed to include the appendix to justify the conclusion and the story, but the story has now not only been justified but greatly expanded. I also considered writing a new appendix, to include what had been done since. However, I decided against that because so much has been accomplished in the years after the book's initial publication, by so many dedicated and professionally competent scientists, that anything I might try to synthesize would be deficient. This, then, is the story of a beginning, and it gives a flavor of the Maine woods in winter, the ravens in the wild, and the people whose passions united in solving a gripping scientific riddle.

PREFACE

As an academic field biologist, I have the duty of finding out about our natural world. I also get privileges, like a sabbatical leave. Most of my colleagues in North America tend to spend their sabbatical years in distant, exotic lands to get to know new organisms or to see new perspectives on old biological puzzles. I went instead to my retreat in Maine because I had seen ravens there behaving in what seemed to me an irrational way, and I wanted to find out why.

It was winter, and at that time the retreat was simply a tiny tarpaper shack, or "camp" in Maine lingo, with a rusty curved stovepipe coming out the side. The stovepipe served wonderfully as an air intake during blizzards. But we'll get to that presently.

The camp, named "Kaflunk" by a previous owner some decades earlier, is in western Maine at the edge of Mount Blue State Park. I grew up near there. It is a good place to grow up, because the woods are endless, and at least then there were no fences and no No Trespassing signs to hem a boy in. Even now I get a sense of lightness and freedom at Kaflunk.

It is situated at the edge of a clearing overlooking a large valley. To get to it, you ascend a steep foot trail through a half mile of forest. The isolation was essential for the observation that I planned to do, but it was not so good for the logistics, such as slogging through three feet of snow while lugging in groceries and dragging up dead sheep and calves and such for the ravens.

I had stayed at the camp before. But it was summer then, and I was working with bumblebees. Bumblebees are highly social animals, and you might expect members of a colony to cooperate with one another. Things don't always turn out as expected, however, mostly because you tend to overlook some essential detail. The bumblebees, it unexpectedly (to me) developed from the detailed data, competed in a scramble competition, with each bee maximizing its foraging returns *individually*. This led to a type of capitalist economy involving costs, tradeoffs, and an optimality solution.

In some of the same fields and woods where I had made the observations on bumblebees, I had often noticed a pair of ravens. I now saw the birds, which had always seemed to me solitary animals, doing something solitary animals are not "supposed" to do: They were sharing valuable food—those who had, it seemed, were giving to those who needed. It was the most left-wing behavior I had ever heard of in a natural system. Furthermore, it did not make sense. (As a biologist interested in how things work, I always look for some evolutionary, self-serving reason for why animals do things, although this is totally apart from the animals' *motives*, and even more removed from what "ought" to be in terms of human behavior.) This time my mind failed to provide a clearly selfish, evolutionary cause for the apparent sharing, and that failure gave me an instant adrenaline rush. I felt that I might not only learn something about ravens, but also something of larger theoretical value.

It is amazing how you can see something every day and yet not *notice* it. There had been many occasions over the years when I had seen groups of ravens feeding at a deer or a moose carcass, as had other people over thousands of years, but nobody had noticed that it was odd. It is odd because by all the canons of common sense and theory a carcass is a very valuable resource (at least to a raven), and any one raven who finds such a food bonanza "should" defend it vigorously if it can, because then it would be well fed for months. If it does not defend the food bonanza, others will eat it, and another may not be found soon. The question almost lunged out: Why were the ravens sharing? What is the underlying pattern that explains the anomaly?

Before starting to try to answer this question, I had read no literature about ravens, and I remained ignorant even after I was well into the work. I wanted to come to my ideas from my own observation, not be guided by what others might expect. Of course, later I read the literature critically. This is also the way I wrote the book. I give a general introduction about ravens and why they interested me, and then jot down my naive observations. Syntheses of the often confusing and contradictory but always fascinating literature on ravens relevant to the observations were added later.

This book is a detective story that tries to solve a puzzle, a hunt for elusive game. It is about searching for clues by watching ravens day after day and sometimes forcing them to yield evidence that would provide a coherent picture of how one little piece of Nature operates. Right from the start I felt that after the solution was found, it would—as is usually the case—seem almost self-evident and then quickly be taken for granted. If it makes very *good* common sense, we say it is self-evident. After it fits into a theory (which is, after all, only formalized common sense), we feel that it could have been predicted.

But the lure is the hunt itself, not the prize. As research biologists, we mount the trophy between the pages of a prominent journal—where it will, we hope, catch the eye of admiring colleagues. But most biologists, hunters, and problem chasers are too busy and too absorbed during the chase to preserve a travelogue of the hunt. Our eyes are close to the ground, and our minds are too absorbed to stop, reflect, and write it down. Maybe this is because, for most biological chasing and sleuthing, there are no clear starting and stopping points.

This study had only a reasonably defined beginning because I'm a neophyte in corvid research, and the project unfolded in a series of field trips to the study site. These trips became natural steps in a set of continuous observations. I kept a record of each step as it happened, so I could see the wrong turns, the right, the relevant, and some of the irrelevant as well.

The text of this book was derived from my field notes. The first set of notes I took gave details such as the precise time that a raven arrived, whether it dipped its wing, made a quork and flew on, and so forth. I recorded the kind of bait, the presence of other birds, and anything else that might possibly be relevant—unfortunately what was really relevant was often not apparent except through hindsight. At the end of the day I read over these notes to extract data that seemed worth accumulating and made a daily summary of what I had hoped to find, what I had expected, and what had actually occurred. It is primarily these second notes that are presented here, with some relevant background information.

Ravens are extraordinarily hard to work with, especially in winter. In my study area the lowlands are dense thickets of white cedar and balsam fir. The steep ridges are mantled in oak, beech, and maple, and the ridges are capped with thick stands of red spruce. The snow that covers them is several feet deep. Temperatures often dip below −30°F. Blizzards are common, and ravens are very rare in comparison with crows and blue jays, their close relatives.

My ravens are also unusually shy, flying off if they see someone stop to look at them even from a distance. Previous work had suggested that they probably range over great distances, possibly hundreds of miles. Their sex cannot be determined from external appearance, and from many reports they are wily and almost impossible to catch.

In short, ravens are near the bottom of the list as a sane choice for a research project. I already knew (although not nearly as well earlier as later) that it would not be an easy project. I would need a tremendous amount of luck, or hard work, or both, to bring it off. Popular belief contends that it is next to impossible to come in contact with ravens, and you cannot hope to learn from your animal unless you gain that contact.

More has probably been written about the raven than about any other bird. But definitive scientific studies are very few, appearing mostly in obscure journals and often in German. Most of the literature consists of notes and anecdotes, and many of the conclusions are false or misleading. Furthermore, much of our "knowledge" is clouded (or illuminated?) by centuries-old myths and folklore, as well as by misidentification. Yet in 1872 the American ornithologist

Edward A. Samuels wrote in *The Birds of New England*: "The habits of this bird have been described so many times, and are so familiar to all, that I will not give them extended notice here." Samuels was wrong. Even now the raven is truly a bird of mystery. I hope here to provide an authoritative book on this bird in the context of solving a scientific puzzle. I have undoubtedly left out many a favorite raven story, and for that I apologize. I was forced to exercise stern judgment about what to include; otherwise it would have taken several volumes to give the bird the full coverage it deserves. Ultimately, the book is about the problem. I was less concerned with summarizing all the facts than with presenting a few new ideas.

The raven project involved a seemingly inordinate, continuous effort that would not have been possible without the generous and dedicated assistance of numerous people who were always interested and cheerful and who made the work fun. I thank Lenny Young and Kate Engel for providing invaluable advice and critical supplies for the telemetry and marking. Billy Adams, Ola Jennersten, George Lisi, James Marden, Brian Mooney, A. Rosenqvist, Charles Sewall, Steve Smith, and Wolfe Wagman all partook in the memorable raven roundups. Gillian Bowser, Denise Dearing, Steve Ressel, Laura Snyder, and Wolfe Wagman were there to help take care of the vociferous raven youngsters when I could not. Wolfe Wagman, Delia Kaye, Leona and Henry DiSotto, Alice and Denise Calaprice, Brent Ybarrondo, O. Jennersten, Elsie Morse, John and Colleen Marzluff, C. Sewall, S. Smith, J. Marden, Dan Mann, Jesse Graham, Billy Adams, Scott Dixon, Stephen Card, Kimberly Frazier, Michele Kruggel, and the Wojcik clan all attended and vigorously participated in the giant raven cage-raising parties, as well as the preceding events. Other logistic support was given by Vernon Adams, Dana Eames, Christel Lehmann, Lee Lipsitz, and Gus Verderber. I thank Dave Hirth, Bernie Gaudette, and Mike Pratt for alerting me to carcasses when they were critically needed. David Capen, Pamela Duell, Lincoln Fairchild, and Peter Marler provided the equipment and expertise that made the sonograms and the critical work on vocalizations possible. David Hirth, Moira Ingle, and Dave Person supplied other equipment and helped with the radio telemetry. I profited greatly from the following people who, through correspondence or conversation, shared their expertise about ravens and other aspects of the project: Skip Ambrose, Pat Balkenberg, Warren Ballard, Peter W. Bergstrom, Kathy Bricker, David Bruggers, Cyril Caldwell, Martha Canning, Peter Cross, Jim Davis, Laurel Duquette, Kate Engel, Frank Gramlich, Eberhard Gwinner, Fred Harrington, Gary Haynes, Doug Heard, Joan Herbers, Henry Hilton, John Hunt, "P. J." Johnson, Lawrence Kilham, Hugh Kirkpatrick, William Krohn, Audrey J. Magoun, Miles Martin, John Marzluff, Fran Maurer, Mark McCollough, L. David Mech, Frank Miller, Karen J. Morris, Frank Oatman, Raymond Pierotti, Paul Sherman, Susan H. Shetterly, Robert Stevenson, Charles Todd, Chuck Trost, and M. L. Wilton. I am grateful to the Psychobiology panel of the U.S. National Science Foundation for having the faith to provide me with a "seed" grant (BNS-8611933) that was indispensable for the project. A Humboldt Award from the Federal Republic of Germany gave me time to write. I gratefully acknowledge the unstinting hospitality of Andreas Bertsch, my host while in Germany. Last but not least, Erika Geiger was able to decipher my illegible handwriting to produce the manuscript, which greatly profited from sound editing by my fellow ravenophiles John Marzluff, Rick Knight, and Alice Calaprice, and by Anne Freedgood who "saw" the book and helped put it all together.

Because of all the "community effort" and good times that went into or have come out of this project, I donate half of the potential profits of sales of this book to the cause of further raven research. Royalties are administered by the University of Vermont under the Raven Research Fund. All contributions will be gratefully accepted and acknowledged.

AN INTRODUCTION

T HIS BOOK is about the common raven, *Corvus corax.**
But what bird, exactly, is this? Everyone knows that the
raven is a large black bird. But depending upon where you
live, there are many species of large black birds. In New
England and other parts of the northeastern United States,
if one wants to distinguish the raven from other large black
birds, one has only to differentiate it from the crow (and
occasionally the turkey vulture). Confusion arises when one
shifts directly to another locality, because the raven belongs
to the crow family, and there are forty-one species of crows
recognized worldwide. Unfortunately, many of the same
species have different common names. For example, the
hooded crow, *Corvus corone,* is also called the Scotch crow,
the Danish crow, the Irish crow, and the gray crow, whereas
the African brown-necked raven, *Corvus ruficollis,* is also
called desert crow, raven, brown crow, and Edith's crow. To
add to the confusion, I know of two scientific papers where
this crow, *Corvus ruficollis* (which weighs about one third as
much as the raven, *Corvus corax*), is referred to as "*Corvus
corax ruficollis,*" as if it were a subspecies of the common
raven. I also know of two other recent scientific papers on
the raven, *Corvus corax,* that are illustrated with a drawing
and a photograph of the American crow, *Corvus brachyrhynchos.*

* *Although the American Ornithological Union has decreed that
the specific names of birds be capitalized, for consistency I yield
here to the more common practice of not capitalizing the names
of animals, including birds.*

17

Here in the Northeast, what we call "the crow" is specifically the American crow, *Corvus brachyrhynchos*. The common raven (also called the northern raven, or Yel, Txamsem, Hemaskus, Tsesketco, and other names by various northern Pacific coastal Indian tribes) is *Corvus corax*, as named in 1758 by Carolus Linnaeus, the Swedish biologist who invented the two-name system of classifying organisms by Latin names. In other parts of the United States and Canada the raven is most likely also *Corvus corax*, although a close relative, the Chihuahuan raven, *C. cryptoleucus*, lives in parts of the Southwest.

In the field the common raven, *C. corax*, can be recognized by its large size (commonly weighs about four times as much as the American crow, and its wingspan is up to four feet wide), its pointed wings (as opposed to the relatively blunt and splayed wings of crows), and especially by its long wedge-shaped tail (most crows have relatively square tails).

Ornithologically, ravens are members of the crow family, the Corvidae, and one can avoid the double meaning of "crow" by referring to ravens as "corvids" rather than crows. The Corvidae include not only the typically large black birds of the genus *Corvus*, but also the brightly colored jays, magpies, and nutcrackers. Typically, the corvids are medium- to large-sized birds with nostrils covered with nasal bristles. (However, pinyon jays lack these bristles, as do adults of the European rook.) Males are sometimes slightly larger than females, but sexual color differences are absent. Corvids typically mate for life, although they often quickly remate when a partner dies. They hide surplus food. Both sexes build the nest and feed the young, but, except for the nutcrackers, only the female incubates. All corvids take both animal and vegetable food, and where they are not persecuted, they often associate with humans.

The corvids belong to the Passeriformes, the evolutionary recent order of songbirds, which includes finches, warblers, woodpeckers, shrikes, vireos, and many others. There has been much debate about how to divide the passerine species into different families. Ornithologist Dean Amadon has suggested that the Corvidae should include the birds of paradise.

Charles G. Sibley, formerly of Yale University, has deduced taxonomic affinities from egg-white proteins and concluded that they are "impressively uniform" among the Corvidae, although the proteins from corvids are apparently more similar to those of shrikes than to those of birds of paradise, the Paradiseidae. More recently Sibley and his colleagues have attempted to ascertain relationships by examining the extent of biochemical binding possible between the DNAs of different species. The more the DNAs of the two species bind together, the more their genetic information content matches (for example, humans and chimpanzees can be shown to share about 98 percent of the same genetic material), and the closer they are related. Although such studies have not been entirely uncontroversial, they show that the corvids are more closely related to birds of paradise than to shrikes.

In general, shrikes and birds of paradise adapted to forest habitat, while the corvid line (except for jays) radiated out to occupy open land. Many of the more recently evolved corvids now forage at least partially on the ground. Some have even adapted to treeless country and to nesting on cliffs. Given the tendency of corvids to be large, intelligent, adaptable, ground-foraging birds independent of trees, it is probably only a slight exaggeration to say that the raven *C. corax* is the ultimate corvid. If so, it is also at the top of the most species-rich and rapidly evolving line of birds. It is the *ne plus ultra* of up-and-coming birds.

Despite all the caveats about the particular corvid or crow-like bird that may be called a "raven" (two species in North America, one in Europe, four in Africa, and three in Australia), in the public consciousness of Europe and America and in most of the extensive literature, *raven* refers to one species only: *Corvus corax*. It is this species that is the primary object of comments and observations in folklore, scientific literature, and this book.

The raven, *C. corax*, occupies an extraordinary geographical and ecological range. It is circumpolar, found even above the Arctic Circle and all the way south to the mountains of Central America. Its ancestral range probably included most of Europe, Asia, and North America. It lives on the frozen tundra and on arctic ice floes, in dense coniferous as well as deciduous forests, in hot deserts, and, more recently, even in some urban areas.

Worldwide, eight subspecies of the raven, *Corvus corax* Linnaeus, have been recognized in Ernst Mayr and James C. Greenway, Jr.'s, authoritative *Check-list of Birds of the World*, although such subspecies recognition is somewhat arbitrary. For example, Malcolm Jollie at Northern Illinois University in DeKalb, Illinois, believes that the many subspecies are not justified (he suggests lumping six of them into one) because there is a great deal of variation in *C. corax*, and the variation is in line with the environmental variation (desert birds are paler, and northern birds are bigger).

The *C. corax sinuatus* race, the western raven, is well recognized, and it differs from *C. c. principalis*, the northern and eastern raven, primarily in being much smaller. But if size alone were a valid taxonomic characteristic, then the measurements accumulated by George Willett working at the Los Angeles County Museum indicate the possibility of still a third race, ranging from the interior valleys of California into Mexico. Besides the apparent size differences in different regions, there is considerable variation even at the same place and time. In one sample of fifty-six ravens from western Maine that I studied in January–February 1986, body mass ranged from 1.05 to 1.53 kilograms, averaging 1.22 kilograms. (For comparison, an American crow weighs approximately 350 grams.) The Maine ravens thus appeared to be similar in size to Alaska ravens, where males averaged 1.38 kilograms. Bill lengths in the Maine ravens averaged 8.18 centimeters and ranged from 7.5 to 9.3 centimeters. Bill depth ranged from 27 to 32 millimeters, averaging 30.5 millimeters.

Whatever its exact size and name, the raven is big, black, and beautiful. Its highly glossed plumage shows iridescent greens, blues, and purples, shining like a black dewdrop in the light. And it dives and rolls like a black thunderbolt out of the sky or speeds along with liquid, gliding strokes. The raven is the paragon of the air, and more. It is assumed to be the brains of the bird world, so its deep, sonorous, penetrating voice demands immediate attention and respect, even though we have little or no idea what it says. It has a greater variety of calls than perhaps any other animal in the world except human beings. It is an imposing bird.

Ravens associate with any animals that kill large game— polar bears, grizzlies, wolves, coyotes, killer whales, and humans. All large-scale northern hunters have their retinues of attending ravens. In the Arctic the Inuit and other native peoples know when the caribou arrive on their migrations by the announcements of the ravens who travel with them and feed on the kills of the wolves along the flanks of the herds.

Early hunters left few records, but it would be surprising indeed if the raven has not been associated with humans for as long as we have inhabited the northern hemisphere. The raven attended our forebears' kills, watched them, and followed them back to the camp under the cliff. The raven's image is represented in the Death of the Birdman scene at Lascaux. Even now, in the far North where humans subsist as hunters, the raven frequents the villages. In his classic *Life Histories of American Birds* (reprinted in 1964), Arthur Cleveland Bent describes ravens in the Aleutian Islands as being "tame as hens." Ravens still associate with human beings by scavenging at garbage dumps, which are the current analog of picked-over "carcasses."

The raven has earned a prominent place in the mythology of northern peoples, in both the Old and New Worlds. According to Nordic legend, Odin, the lord of the gods, kept a pair of ravens perched on his shoulders. They were Hugin (Thought) and Munin (Memory), and he sent them out at dawn to reconnoiter to the ends of the earth. At night they returned and whispered into his ear the secrets they had learned. Odin chose his messengers well, because no bird is a better long-distance flyer or more sharp-eyed, alert, and loquacious than the raven. (Can a raven miss anything? Can it keep a secret?) Odin, with his universal knowledge, then advised the other Norse gods. In ancient Ireland, future events were divined from the calls of the raven, and even now the Irish phrase "raven's knowledge" means to see and to know all.

The raven's flight activity and loquaciousness were un-

doubtedly at fever pitch when the Vikings went into battle.
The raven was the battle bird, and Viking warriors reputedly
carried a sacred raven standard, as did William the Con-
queror. The Vikings welcomed the company of ravens, but
undoubtedly the association was based on the ravens' own
practicality or adaptability. They followed the Vikings for
the same reason they now follow the wolves on caribou migrations: to find food.

The Vikings revered the raven, but those whom they
raided feared the big black birds. Ravens were rightly associated with death, and not just in the context of Viking raid-
ers. In Old English literature there are repeated references to
the raven at the scene of battle, as in the great heroic poem
of *Judith* (lines 205–211), where the raven is referred to as
the lank one, the dewy-feathered one, and so on: "[The battle noise] rejoiced the lank one, the wolf in the forest, and
the dark raven, the slaughter-greedy bird. Both knew that the
warriors intended to provide for them a feast of doomed
warriors; and behind them flew the eagle eager for prey, and
the dewy-feathered one, the dark-coated one; he sang a battle
song, the horny-beaked one."

And here are lines 60–63 from *The Battle of Brunanburh*,
at the conclusion of the poem, where the Vikings are van-
quished and retire to Ireland and the Saxons return vic-
torious: "They left behind them, to enjoy feasting on the
corpses, the dark-coated one, the swart raven, with the horny
beak. . . ." Similar scenes are evoked in the early eighth-
century Old English epic *Beowulf* (lines 3021–27): "Therefore shall the spear on many a cold morning be brandished
in the land, lifted up by the hand; not at all shall the sound
of the harp wake the warrior, but the black raven, eager for
the doomed ones, as he shall say much to the eagle of what
success he had at feeding, when he, with the wolf, plundered
the corpses."

The association between ravens and death led to the assumption that the birds could predict death, and the ravens'
hoarse croaking was thought to be a prophecy of calamity
all over Europe and parts of Africa and Asia. To be sure,
ravens will call *after* a death that interests them. And it is
not unlikely that they could also correctly foresee impending

deaths (though probably not an individual death). In medieval times a traveler along a country road may well have heard the sonorous calls before coming to a crossroads where malefactors were strung up to serve as an example. The scene is even now preserved in our language: "Ravenstone" is an old English term for a place of execution.

The raven was probably disreputable not only because it ate carrion, but also because it reputedly did not feed its young properly (young ravens are indeed conspicuously noisy when calling to be fed). In general, raven came to be synonymous with "sinner," despite biblical allegations that ravens fed holy hermits. We read in 1 Kings 17:6 that Elijah had prophesied a drought in Israel, and in so doing stirred up the wrath of King Ahab and Queen Jezebel. God's message was: "Depart from here and turn eastward, hide yourself by the brook Cherith, that is east of Jordan. You shall drink from the brook, and I have commanded ravens to feed you there." According to the Bible, the ravens brought Elijah food. (Did they bring him food because by listening he discovered where they were feeding on a kill? If so, then this should be a profound lesson for us to use our rationality to interpret Nature correctly.)

William Shakespeare, true to the tradition of his time, treated the raven as a symbol of evil and destruction. In *Macbeth* the raven "croaks the evil entrance," and in *Othello* the raven flies "o'er the infected house." In German an evil person who ought to be hanged is still called a *"Raben-aas"* (Raven carrion). In the Middle East, ravens were perhaps less feared as omens of death, but they were not in good standing. According to Jewish folklore, the raven earned considerable disfavor for repeatedly violating the decree against lovemaking on the Ark. The raven was also the first one to be sent from the Ark to look for land. It did not return, possibly because it found floating corpses to feed from, and Noah then sent the dove.

According to E. A. Armstrong in his 1970 book, *The Folklore of Birds*, the tradition of using birds to find land is an old one among mariners. Babylonians used the raven, and Pliny the Elder, the old Roman naturalist, states that the mariners of Taprobane (Ceylon) carried ravens in their ships

and set their course by following them. The Vikings used
them as well. In A.D. 874 Floki, a Norwegian explorer, set
out to find the large island to the west that had been discovered some ten years earlier by a Swede named Gardar.
According to the *Saga of Floki*, Floki took three ravens. The
first one he released flew back to Norway. The second, released later, saw no land after circling high, and it came
back to the ship. Finally, the third flew west and did not
return, presumably because it had found land in that direction. According to the saga, Floki followed this one, and that
is how the Vikings discovered the southeast coast of Iceland,
where ravens are revered to the present day (except in areas
where the eiderdown is commercially gathered, since ravens
destroy the ducks' eggs and chicks).

Perhaps nowhere are ravens such a commanding presence
as in the New World. In numerous Native American cul-
tures the raven is both a creator and a folk hero. In his 1983
book, *Make Prayers to the Raven*, about the Koyukon natives, Richard K. Nelson writes:

> Ravens are a part of most days in the boreal wildlands,
> flapping determinedly towards some unknown destina-
> tion, performing acrobatic follies in pairs or trios, croak-
> ing loudly somewhere in the distance. They remain in
> the north summer and winter, going about their dubious
> affairs regardless of heat or bitter cold. And they are
> everywhere, from dense river forests to broad muskegs
> and meadows, even to the tundra mountains, where
> they play and circle on the rushing updrafts. Whatever
> else ravens may be, they are indeed successful. But then,
> who should know better how to live on the land than its
> own designer?

In the 1988 book *Moose* by Michio Hoshino, the author
quotes Catherine Attla, an Athapaskan Indian. Attla was
talking about moose hunting:

> Sometimes people call on Raven for help. One of the
> things we say to Raven while we hunt is "Tseek'aal,
> sits'a nohaaltee'ogh," which means "Grandpa, drop a
> pack to me." If the bird caws and rolls, it is a sign of
> good luck. Raven is protected because it is said he helped

shape the world. That is why the one who raised me used to tell more Raven stories than any others. He was a medicine man, and he was familiar with Raven power. People also talk to Raven when they see it out in the woods, especially when they are alone. They talk to Raven the same way we pray to God.

According to the various tribes—Tsimishian, Haida, Bella Bella, Tlingit, and Kwaikiutl—of the Pacific Northwest and including the Koyukons in Alaska, raven is the god who created the earth, the moon, the sun, the stars, and people. Raven myths are legion and too numerous to recount. In 1909 Smithsonian ethnologist John R. Swanton published twenty-eight in *Tlingit Myths and Texts*, after a four-month field trip in 1904 to the Tlingit Indians on the northwest coast at Sitka and Wrangall, Alaska. Although the raven was never evil in Native American mythologies, he was often a rascal. For example, raven created mosquitoes to plague people. In Inuit legend, raven created light by flinging glittering mica chips into the sky, and the Milky Way marks this track of mica across the heaven. To raven the god, human beings are part of the menagerie he has created for his own amusement. First he created humans out of rock, but that made them too durable, so he used dust to make them become mortal, as they remain today. In the original perfect world that he created, fat grew on trees, and rivers flowed both uphill and down. But this also made things too cushy for humans. So he changed the fat to fungus and made rivers run downhill only. He also devised an assortment of other difficulties for man in his role of mischief-maker, clown, and god.

Koyukon shamans as well as those in more southern tribes still invoke raven's power to try to scare away sickness by mimicking his cawing, spreading their arms like wings, and hopping up and down on both feet.

In ancient times, North American Indians, Chinese, Greeks, Siberians, and Scandinavians believed that raven controlled or affected the weather, and on a recent canoe trip down the length of the Naotak River in northwestern Alaska, when I remarked to two Eskimo park rangers about the constant rain, they explained that rain is caused when someone kills a raven. (Incredible as it may seem, I found a dead raven

near a trapper's cabin the next day. The bird had been dead
for at least a week. There was no telling if someone had
killed it, though.)

The ancient myths and legends about ravens are not just
interesting esoterica. They determine attitudes that affect
the birds' distribution, and I believe they may even influence
the food-sharing behavior that we will discuss later.

Wherever the raven is a god, it enjoys a charmed life
around humans. Among the Koyukons and other northern
natives, it is taboo to kill a raven, and if one gets caught in a
trap, it must be released alive while the trapper tells the bird
that it was not meant to be caught. This taboo seems to be
ancient. Henry B. Collins, during his archeological excavations at Cape Kialegak at the southeastern end of St. Lawrence Island, which has been occupied since A.D. 900, uncovered the remains of forty-five species of birds killed by
the Eskimos. He concluded: "The absence of raven bones
showed that in prehistoric time, just as today, the raven was
regarded as sacred by the Eskimos and never killed."

Ravens congregate in or near Eskimo villages all over the
north, and in towns like Yellowknife and Inuvik they grab
up refuse, raid groceries left unattended on pick-up trucks,
and steal the sled dogs' food. In Iceland, where there is still
ancient Nordic respect for ravens, the birds are quite tame.
In parts of the western United States the raven has accom-
modated to humans, as it has to almost everything else.

In the eastern United States, however, the raven is still a
symbol of the wilderness, and over large areas of Europe it
has been eradicated by poisoning bait, shooting, and destroying nests. A pioneer raven researcher in Germany, Johannes
Gothe, notes that, in the dukedom of Mecklenburg-Schwerin
alone, 10,440 ravens were listed as shot in the forty-one year
period 1834–1875.

Similar persecution followed the immigrants to the New
World. According to E. H. Furbush in his 1927 book, *Birds
of Massachusetts and the Other New England States*, after
European settlement in the East, the raven "soon became
known as a killer of sickly sheep and new-born lambs, and
the settlers waged a relentless warfare upon it." The English
and German settlers loathed and feared the birds, and they

were perhaps overeager to attribute the death of a sickly sheep to a raven, particularly if they found ravens feeding on the carcass.

The persecution continued in the West. In his book *Hunting and Trading on the Great Plains: 1859–1875*, James R. Mead also comments on the association (then often fatal) between ravens and carnivores and their prey. Mead recounts how in the fall of 1859 Chief Shingawassa of the Kaw (Kansas) Indians camped in the timber in the back of his ranch. They stayed through the winter, taking and trading furs for goods (coffee, sugar, flour, and tobacco). He writes (p. 73):

> Our method of killing wolves was to shoot down two or three old bull buffaloes. . . . We would let the buffalo lie one night in order to attract the wolves. The next night, just before dusk, we would go and scatter poisoned bait about the carcasses, each bait containing about one thirtieth part of a dram of strychnine. The reason we put out our baits after sunset was an account of thousands of ravens that seemed to live with the buffalo, and which were confined exclusively to country occupied by them. They would come back and pick the baits if put out before dark, so that instead of killing wolves, we would find we had a whole field of ravens killed.

He continues (p. 74):

> These ravens did not nest in that section of the country—at least I never saw their nests in my travels. The buffalo, the gray wolves, and the ravens—companions in life—mingled their bones when swift destruction overtook them. The buffalo were killed by the bullets of the hunters, the wolves were killed with strychnine for their furs, and the ravens died from eating the poisoned carcasses of both, so that they all became practically extinct at about the same time.

Persecution of crows and ravens continued on to this century and to the present day. Crows are convenient scapegoats for crop failures, so they are relentlessly persecuted as "vermin." (A classmate of mine at the University of Maine paid his college expenses by working for the Fish & Wildlife De-

partment in northern Maine. His job: killing ravens who
were eating potatoes and supposedly "spreading disease.")

To the average person who pays $7.00 for a hunting li-
cense, there is no distinction between crows and ravens. An
illustration of how we treat "crows" is encapsulated in an
incident at a communal crow roost (ravens sometimes roost
together with crows and other corvids) in Illinois in the
1940s. The trees where the birds slept at night were festooned
with a thousand hand grenades, which, when detonated at
night, left the ground spattered with about one hundred
thousand dead and dying birds. In the 1960s Bert Popowski's
Varmint and Crow Hunter's Bible was still able to recom-
mend crows to hunters to "take up the off-season slack in
available targets."

In 1972 the Migratory Bird Treaty Act was amended to
include ravens and crows, but most states list "crows" as a
legal "game" bird. They are treated differently from other
game birds, however: there is no bag limit, and they are fair
game not only in the fall but also in the spring, during the
breeding season, when they are feeding their young. One
brochure, printed by a state Department of Wildlife Con-
servation, makes specific recommendations on crow shooting
(at least it is honest and does not mention *hunting* in the
title of the brochure) at a well-known roost. It reminds the
reader that there are no bag limits on crows. Shooting hours
are: "Daylight to dark daily. Early morning and late afternoon best times." Two other useful hints: "Find their flyways
for best shooting," and "Bits of tin or tin cans may be hung
200 yards away from the opposing side of the flyway to drive
the crows toward the blind." The various kinds of calls to use
or not use are also described.

A January 1985 article in *Fur-Fish-Game* magazine makes
no pretense that shooting crows serves any useful purpose.
It simply advises the shooter to "discard dead crows in a location specified by the landowner. Don't just leave them laying
in a pile in a field or behind the barn." The article concludes
that crow "hunting" is "a sure cure for cabin fever."

Incongruously, it is illegal to tamper with the *nest* of a
crow, and any researcher who wants to study them must get

state and federal permits. Having a pet crow is strictly against the law. (Why? Much can be learned from a live crow, little from a thousand dead ones.) Why should anyone have compunctions about breaking this law, when doing so to get to know an animal from the wild might do much to raise ecological consciousness?

Both crows and ravens have been persecuted in the eastern United States, but ravens until recently declined while crows greatly increased. This may be partly because ravens' annually reused nests (or nest sites) are very easy to find and destroy. But there are other reasons as well. In former times the raven was found throughout North America, but it disappeared from the prairies with the passing of the buffalo and the wolves. Poisoned baits took their toll. The ravens' food base, the buffalo, was also removed, while the crows' food base, which is associated with agriculture, increased. By the beginning of the nineteenth century the raven was gone from most of New England. Henry David Thoreau does not mention the bird in writing about his three trips to Maine, and in 1872 Edward A. Samuels in *Birds of New England and Adjacent States* observed that the raven is an "extremely rare resident in New England; but it occasionally rears its young on the Island of Grand Menan, off the north-east coast of Maine—on all but inaccessible cliffs." In 1903 the ornithologist Thomas Nuttall wrote: "Of late years the Raven has almost forsaken New England," and several decades later Arthur Cleveland Bent concluded that the raven was "uncommon or rare over most of its range in the United States," with the New England states being "outside the normal range." Similarly, in 1912 Walter B. Barrows wrote that in Michigan the raven is a shy bird who "disappears when settlement advances."

It has not always been so. Ravens at one time existed in great numbers even within the city of London. They acted as the city garbage crew, and in the seventeenth century a flock of them alerted Charles II's guards when Oliver Cromwell attempted a raid.

I gleaned additional information on London ravens by writing to the Tower of London, where I knew ravens are

still kept. The following are excerpts from a letter dated 14
February 1989 that I received from John Wilmington Bem,
the Yeoman Raven Master at the Tower:

It was in Charles II's time, according to tradition, that
the ravens were kept and looked after properly at the
Tower. Prior to this there were ravens all around the
area, especially Bermondsey, and after the Great Fire of
London in 1666 they gorged themselves on the bodies
which had been left because the authorities couldn't cope
with clearing the debris. They multiplied and became
such a nuisance that the residents petitioned the King to
get rid of them all. However, a soothsayer advised the
King that if he removed all the ravens from the Tower
a great disaster would befall England and his Royal Pal-
ace would crumble into dust. The King, not wanting to
tempt fate, decided to keep six ravens and appoint a
Keeper, of which I am the last up to now, but hopefully
there will be more to follow me.

When a raven dies, he is buried in the Moat near
Traitor's Gate and his name recorded there. If there are
no guest ravens at this time, I ring various people to find
out if there is a lame bird available, or one which has
been raised as a pet and is now unmanageable. Gwylum,
our latest arrival, came from the Welsh Mountain Zoo.
I clip them by cutting the pinion feather on one wing.

The normal establishment of ravens, said to have been
brought in on the orders of King Charles II, is six, but
from time to time, at the Governor's discretion, we also
have two guest ravens. Each raven needs quite a lot of
room because, being unable to fly, they are hopping
around on the ground all the time, so six to eight is a
manageable figure. We have eight at the moment: Rhys,
Charlie, Hughin, Larry, Hardy, Gwylum, Katie and
Cedric. Last year there were three mated pairs which
has never happened before, and this caused untold trouble as they became very aggressive.

They are allowed out of their cages as soon as it's
light and fed their daily rations, which is finely minced
meat with dog biscuits and kitchen scraps. They are
slightly overfed which makes them docile and less likely
to get annoyed with the visitors. They are put back to
bed as it gets dark, and each has its own apartment with

straw to keep it warm and louvred doors to shut out the light. They will come when I whistle, but when I'm not here other people find it difficult to persuade them back into their cage. Each raven has its own characteristics; some are vicious, some love you to cuddle and pet them. They love attention, and I have to be careful not to have favourites as they get jealous!

Ravens do not readily breed at the Tower. They have mated and produced eggs in the past, but the eggs have always been destroyed after a few days. It may be that they are disturbed by the visiting public and have too little privacy. There is also often building work going on in the grounds which may put them off. We may try once more this year and, if we get any eggs, we will try to incubate them ourselves so that the chicks will be born in the Tower of London, which is what we are trying to achieve.

Ravens may have endeared themselves to many people in Great Britain, but shortly after Charles II's time they were apparently persecuted, nevertheless. Robert Smith, who in the third edition (in 1786) of his book *Universal Directory for Destroying Rats, and Other Kinds of Four-footed and Winged Varmin* described himself as "Late Rat-Catcher of London," reported his tricks for catching ravens and declared, "I have caught great numbers of them in a day." Ravens were trusting then. Just fifty miles southwest of London in the village of Selborne, a pair of ravens had nested for years on an ancient oak that had become known as "the Raven tree." The Reverend Gilbert White, famous for his meticulous observations on nature recorded in *The Natural History of Selborne* (first printed in 1788), also remarked on them. The ravens nested high on a jutting bulge of the oak, and generations of village boys had tried to reach the aerie, but none could skirt the bulge, and so each year "the ravens built on, nest upon nest, in perfect security." Finally, the oak was cut to build a bridge in London. White writes:

The saw was applied to the butt, the wedges were inserted to the opening, the wood echoed to the heavy blows of the beetle or mallet, the tree nodded to its fall; but still the [raven] dam sat on. At last, when it gave

way, the bird was flung from her nest; and though her
parental affection deserved a better fate, was whipped
down by the twigs, which brought her dead to the ground.

Although ravens were obviously no longer protected, the
king's decree to keep ravens was held to the letter of the law.
As we have seen, six to eight ravens are still kept (as cap-
tives) in the Tower of London, but the spirit of the decree
has clearly been violated.

There are no more free wild ravens in or around London.
But crows and other corvids are found in other towns all over
the world. In the 1950s, ravens started walking the city
streets of towns in the Saskatchewan prairies during the
winter. In Whitehorse, in the Yukon Territory, they are
welcome birds-about-town. They are now even receiving official recognition in Canada. Bill No. 12 of the Twenty-Sixth
Legislative Assembly of the Yukon Territory passed the
Raven Act on June 14, 1985, making the raven the official
Territory Bird. (To achieve this required a petition circulated
by the "Raven Lady," "P. J." Johnson, that was signed by
fifteen hundred "raven maniacs.")

No other bird in the world has a wider distribution or
shows more adaptability than the raven. It is equally at home
following polar bears to their kills in the High Arctic, tagging along with wolf packs in the Canadian taiga, catching
lizards at over 120°F in Death Valley, or flying over the
highest mountain peaks in Tibet and in North and Central
America. I have seen ravens jump into the garbage bins in
back of Dunkin' Donuts in Flagstaff, Arizona, and forage for
roadkills by flying along the highways in Maine and the
Mojave Desert. I have seen a raven's nest on a cliff in the
Truelove Lowland above the Arctic Circle, on high-rise buildings in Los Angeles, and on the steeple of St. Mary's church
in Flagstaff above a parking lot. Ravens nest on cottonwood
trees in the Grand Tetons, on telegraph poles in New Mexico,
and on white pine trees in Maine, as well as on crumbling
rocks a few feet above the water on the Naotok River in
Alaska. Recently they have even been observed nesting on
interstate highway overpasses and billboards. Ravens are at
home everywhere. They have only one enemy: humans.

Despite relatively recent persecution, the raven has been making a dramatic comeback in New England as well as in many other areas. It is likely that its range is still expanding. Ravens came to central and western Maine thirty years ago, apparently close on the heels of the invading coyotes. There were only isolated reports of ravens in Vermont before 1960, but since 1972 they have increased dramatically until they are now found over most of Vermont, Maine, and New Hampshire. Undoubtedly they will continue to spread south in the next few decades. But in the East they are still very shy birds, seldom seen close to towns. They breed in the deep forest and fly away immediately if anyone comes close. They are very hard to get to know.

My acquaintance with this elusive bird began when my family was living as displaced persons in a one-room cabin, deep in a German forest preserve, at the end of World War II. That, at least, is when I first came to love corvids. I was not yet ten years old, and I did not have many toys, but I had the best entertainment and companionship a young boy can have: a pet crow. I raised it from a nestling, and since then I have had pet crows and ravens on many occasions. As some people need to have a cat or a dog, I need to have a corvid.

At that time, between 1944 and 1950, life was one continuous adventure for us. One time my sister Marianne and I were walking along the sandy road through a thicket of spruces on our way to the village school. We were afraid of stags and wild boar, and when we heard a deep croaking and saw great black birds erupting out of the thicket, it did not help matters. Naturally we told our parents, and Papa knew the meaning of what we had seen. Our situation was like Elijah's in the wilderness, and indeed the ravens brought us food, but only because we heeded his "message," which was: "Food here." It was a boar. When fried, it was the most delicious thing we had eaten for a very long time.

I did not see a raven again until the late fifties, at the same site in western Maine that is the focus of the observations described in this book. I was with Phil, one of the best woodsmen and fishermen I have ever known, who did his best to

educate me to be a true Mainer. I don't know if he succeeded
to his own satisfaction, but he certainly had a big impact on
me. On this particular day, we were up to our usual fall activity: deer hunting. Nothing very *exciting* happened on
these forays most of the time—we never seemed to shoot a
deer—but we kept our eyes and ears wide open. This time I
was rewarded by hearing the croaking of a raven—and then
I saw it swoop over the ridge, heading for Mount Tumbledown. Ravens in *these* woods! I was electrified. Might they
nest here, too? I kept the question in mind, because I was
also an avid egg collector, and I had never seen a raven's
nest. I could not even imagine the thrill I might feel if I ever
ascended to peer into a raven aerie and saw the eggs, presumably, like those of most other corvids, greenish with splotches
of black and gray.

At the time, I was working in the kitchen at Kamp Ka-
wanhee on the shores of Lake Webb at the foot of Mount
Tumbledown, washing dishes for $15 a week. Every morning
near dawn in June I heard the raucous begging of the young
ravens that flew behind their parents as they foraged along
the lake shore. I looked for the nest in the tall white pines
along the lake, but I did not find it. Nor did I find a nest for
many years. But I kept ravens on my mind, wondering about
them. Did they nest only on inaccessible cliffs in the nearby
mountains? Did they nest in dense thickets of spruce near the
top of nearby Mount Blue or Mount Bald, or did they prefer
the balsam fir trees along the banks of the brooks?

It was in late March about ten years later that I finally
stumbled on a nest. The snow was still several feet deep, and
the crust on top of it only partially supported my weight as
I worked my way toward the opposite shore of Hills Pond.
The smell of spring was in the air. But what burst upon me
that day was the loud penetrating call of a raven. I had never
heard a raven so close. The quorks were so powerful they
suffused the woods and overshadowed all other sounds. Their
meaning was clear to me: A nest was near.

I searched near the pond and then walked along a low
ridge nestled among the mountains. On one side, red spruces
intermingled with a few large white pines and a few white

birches. If you had a nest in one of those pines, you could look over the full expanse of the frozen lake. If you looked up to the slopes of Mount Bald, you could see the hardwoods grade to red spruce to bare rock on top. Perhaps you could even see down to Lake Webb. On the other side, you could see up the valley of Alder Stream, with Mount Blue in the distance. And it was here, on one of the pines, that I found the nest.

Subsequently I have seen sixteen more nest sites in central Maine and Vermont. Three were on cliffs, and thirteen were on pine trees. Interestingly, Gothe, in his 1961 detailed studies of ravens' nest-site selection in northern Germany, found seventy-one of seventy-three active nests on old *beech* trees and only one in a pine, even though there were extensive pine and other coniferous forests. There are beech forests in central Maine, but to my knowledge no raven's nest has ever been found on a beech tree here. Given that ravens nest on and even in abandoned buildings, on telegraph and high tension lines, and on church steeples, their region-specific nest-site choice is puzzling. It probably has to do with tradition.

A week or so later, when the bird flew from the nest as I got near, I knew she had eggs. There had been a big snow-storm, and the woods looked like deep winter. But in May I saw and heard the big black youngsters at the edge of the nest.

The following year I came in February, but there was no nest. It had blown down during the winter in a storm. The grayed twig ends of the piled sticks told me that the fallen nest was an old one. (Ravens in this region *break* all the twigs—mostly poplar—from the trees for their nests.) I came back a month later and found *freshly* broken twigs, up to three-quarters of an inch thick, on the snow, and a new nest in the tree. It had been rebuilt in the same pine tree, in exactly the same spot. I have since seen this repeated by other ravens, with an intervening year when the nest may be at another, nearby location.

The Hills Pond pair also came back in other years (building up to a half mile away, following two nest failures due to natural causes), and they were to play a central role in my subsequent studies.

Although my original interest in finding a raven's nest had been to get a set of raven's eggs, I never did take any. My egg-collecting phase passed. I kept a few shells, but mostly what remained was the distillate—an interest in the birds' behavior and ecology. I came to see the nest for its own sake, and I hoped nothing would disturb the pair and make them move elsewhere. (Similarly, over the years, I lost some of my excitement at the chase after wild honeybee trees and got involved in studies of bumblebee foraging, instead.)

Foraging behavior—how animals make a living—seems to me to be a pivotal aspect of life. An animal is successful only insofar as it can procure resources from the environment and convert them to more of itself. This, of course, involves maintaining a constant internal environment, avoiding becoming a resource for some other organism and reproducing (i.e., converting the hard-won resources into copies of itself). Many of the latter functions are sporadic or seasonal, but getting the resources to maintain life is an almost constant endeavor. Bumblebees have evolved a fantastic set of behaviors for maintaining energy balance in the summer, even though flowers are everywhere. What I wondered now was how much more fantastic these ravens might be, since they stayed here all winter long when little food was regularly available to sustain them.

If all animals were alike and did exactly the same things, they would soon become boring as scientific subjects. The underlying principles often end up being rather simple, and once you "see" them, they lose their sparkle. They become merely part of your underlying assumptions about how Nature operates. The more general the principle, the more tedious it tends to become. It is variety that excites. And Nature is inordinately more intricate than the human mind can even begin to perceive.

The reason corvids are so exciting is because, while they are similar in many components of their behavior, these components have been "stretched," that is, modified or put together in different ways in the different species to result in entirely new strategies, all of which serve the same general principles of feeding economy at different resource distributions.

Resource distribution often drives social systems. One of the most fascinating and well-known examples is that of the Florida scrub jay, *Aphelocoma coerulescens*, as studied by the zoologists Glen E. Woolfenden of the University of South Florida in Tampa and Woolfenden and John W. Fitzpatrick of the Field Museum in Chicago. Florida is a nice place to live if you are a bird, because the weather is tolerable year round, and there are few sudden environmental changes that cause die-offs and vacancies. Real estate is scarce, however. In Florida a young jay living in the low scrub-oak has relatively little chance of finding another unoccupied spot if it leaves home. What is does very often, instead, is to stay with its parents and make itself useful by defending the family turf and helping to rear subsequent broods there. Thus, several birds besides the parents can usually be seen attending the nest. One of them might later be lucky and inherit the territory. Strangely, this communal breeding has not evolved among members of the same species of jay that live in California, yet as Jerram L. Brown of the State University of New York at Albany demonstrated, it has evolved in an entirely different species, the Mexican jay, *A. ultramarina*.

Russell Balda and his associates at the Northern Arizona University at Flagstaff found a contrasting social system in the pinyon jay, *Gymnorhinus cyanocephalus*. These jays specialize in the seed crops of pines, such as those of the pinyon pine, *Pinus edulis*. Like many other plants having highly prized, nutritious seeds, the pinyon has evolved a famine-and-feast strategy which ensures that some seeds will survive the onslaught of seed eaters. In any one area all the trees may for a number of years produce no seeds. This keeps the seed-eater population low, and when the pines suddenly produce a bumper crop, they satisfy local appetites and still have seeds left over that grow into new trees.

Living off a resource evolutionarily designed to saturate the local environment, the pinyon jays face a more specialized reality than many other birds whose economic incomes

depend on the amount of real estate they hold. With plenty
of locally available food, there is no need for the birds to keep
individually held large territories. Instead, the problem is to
escape local bounds and find widely dispersed, rich feeding
areas. Pinyon jays breed in loose colonies in such rich feeding areas. This makes it possible for them to enjoy some of
the advantages that group living affords: Together they can
repel powerful predators such as hawks, and they can forage
as a group, leaving the guarding of the young to a few baby-
sitters. Information sharing about food-rich areas is possible,
because it costs little in terms of food given up, and since the
flock is a social unit, the sharers are also apt to be friends and
relatives. Furthermore, while they are foraging as a group,
the birds have more eyes to detect danger.

In corvids, specific behavioral traits as well as the social
system may be modified for unique resource distributions.
Caching food away is one example. This is a neat way to
make use of food bonanzas that the animal cannot eat in one
sitting. Just as an owl that catches a hare can get more than
just one meal out of it, crows cache extra meat if they come
upon a bonanza. To most crows, this is not a major or central
part of their life strategy. It is simply something they do as
opportunity arises. Yet, depending on the environment,
caching assumes greater or lesser importance in different
species and may be modified by evolution. Caching behavior
becomes more advantageous when the food is less likely to
spoil. And the concentration on such food increases as caching evolves, in a self-reinforcing spiral.

There are at least three different species of corvids for
whom caching is the major key to survival. One is the gray
jay, *Perisoreus canadensis*, or "whiskey jack," a sometimes
ridiculously tame bird found across the taiga and in mainly
coniferous forests of North America (although I have also
seen it nest in willow thickets along streams in the High
Arctic tundra). The gray jay nests in late winter, long before
the sometimes deep snows have melted. Thus its greatest food
requirements, when it is rearing young, occur when the environment has the least food to offer. However, the coldness
of its habitat becomes an advantage: Food does not spoil, and

the bird has evolved enlarged salivary glands that produce a sticky saliva with which it glues food caches above ground where the deep snows cannot obliterate them.

The Clarke's nutcracker, *Nucifraga columbiana*, has hit on a similar solution not only for surviving the energy crunch of winter, but for breeding then, too. Apparently nearly its entire food supply for its late-winter nesting is derived from cached food. The bird has developed a special sublingual pouch in which it can stuff up to ninety-five pine seeds and carry them as far as 22 kilometers to its storage areas on south-facing slopes. Nutcrackers are largely mountain birds. It can be windy in the mountains in winter, and it isn't easy to glue pine seeds onto branches. What the bird does instead is to forage in the lowlands in the fall and then fly up to high elevations, making use of the updrafts as a sort of "elevator," to deposit the seeds in wind-blown areas where it will have an easier job digging them back up when it needs them months later. The amazing capacity of nutcrackers to memorize apparently thousands of different individual caches is a marvel that has long been of interest to naturalists and is even now under intense research by Russell Balda at Northern Arizona University at Flagstaff and by Alan Kamil at the University of Massachusetts at Amherst.

Pinyon jays also have, as one might expect, polished up their caching routine. Their pine seeds are stored close to the standing trunks of trees, where the sun melts the snow faster, and throughout the winter and in early spring they retrieve them. But they do not depend entirely on caching. Pinyon jays are narrow-winged, making them good long-distance flyers, a second mechanism that helps them foil the scarcity of their favorite food.

There are many other adaptations that have enabled corvids to live, reproduce, and feed themselves and their young. And learning about this background of prior research made ravens all the more exciting to me. If ravens indeed specialize on carcasses, here is a unique kind of food with a rather extreme geometry of distribution and abundance. If any corvid should have unique behavioral adaptations to help live off this resource, it should be the raven.

The background by itself was not, however, what led me
to study ravens; it was only the fuel. The spark that ignited
my interest and started this study came in October 1984 during my sabbatical leave from the University of Vermont,
when I watched a crowd of ravens in Maine near where I
had seen ravens at intervals for about three decades. Some
of the next chapters are the journal I have kept since then,
and others are reviews of the scientific literature, ideas, and
research results.

RAVENS AT A MOOSE

*Much I marvelled this ungainly fowl to hear
discourse so plainly,
Though its answer little meaning—little
relevancy bore. . . .*

—EDGAR ALLAN POE,
The Raven

OCTOBER 28, 1984. Like innumerable other glacially scraped ridges in western Maine, Gammon Ridge has a boulder-strewn top overgrown with cushions of soft moss. The moss is brilliant green in the shade of the dark green, almost black red spruces. White-tailed deer bed down here where they have a view into the red oak, sugar and red maple, and beech forest below.

Trails used by generations of deer have been worn into the sides of the steep ridge. All of the smooth trunks of the large beech trees are scarred with the claw marks of black bear, and the top branches of many are broken and pulled in tangles that look like untidy hawks' nests. These are also the work of the bears, who climb the trees, pull the branches toward them, and feast on the unripe nuts before they fall. Young red maples growing in the bogs show long vertical grooves where moose have chiseled off bark with upward sweeps of their sharp incisors.

In late fall you see small trees and underbrush with the

rub-marks of deer, and the rutting bucks also leave hoof-prints and pawed ground along trails. In the years when the beech trees have nuts, the bears work the leaf litter under the trees into furrows. Hordes of blue jays, evening gros-beaks, chipmunks, red squirrels, mice, and even woodpeckers feed on the remaining nuts. (But strangely no crows or ravens seem to feed on them.) And after the first snow has fallen, the tracks of fisher, bear, and coyote course over the soft white landscape among the gray beech trunks and trail off into the black green spruces.

Now, in late October, the colorful leaves are already down. The migrant birds have left, and the woods are quiet. But the stillness is punctuated here and there by the churring of a red squirrel, the caw of a crow, the scream of a blue jay, and—if you are lucky—the croak of the raven.

The raven's deep resonant "quork, quork, quork" com-mands attention. The calls can be heard over a mile away. And when you hear them, you imagine, or see, somewhere in the distance, a big black bird ascending the ridges with ease, shooting down a valley like a black thunderbolt, and with effortless grace ascending up over the next ridge. These are the north woods, and the raven is their symbol. I love these woods, and my attention is directed like iron filings to a magnet to anything that hints of "raven."

The dampness this foggy morning makes the moss lumi-nous. The recently fallen leaves are turning brown and smell nutty. Their softness muffles my footsteps as I wander with my senses alert.

There—those are raven calls! They are perhaps a half mile away on the other side of Gammon Ridge by the new log-ging area, and I approach them at once, drawn by an irresist-ible force. There is no reason. I just have to go. From past experience I already know that the birds are at an animal carcass. The only question is, what have they found this time?

As I begin to descend from the ridge, one of the huge birds flies up, its heavy wingbeats rending the air in sharp, loud swooshes. Another, another—and then fifteen or more scatter from the same place on the ground among the birches in

front of me. They vanish like black ghosts in the fog among
the white birch trunks. Silence returns.

The ravens had been feasting on the remains of a moose
(one left by a poacher that had been all but totally covered
up by brush and logs). The remaining meat is still fresh.
I remove the hide, cut off a few chunks of meat, make a
small fire up on the ledges by the spruces, and prepare my
simple meal. I wait to watch the birds. There is no greater
pleasure than eating roasted moose while resting under a
spruce and contemplating ravens. I have time, and I listen.
In a half hour only four or five ravens return to the area,
and they stay far away. But they call.

As I listen closely, I notice an endless variation on the
basic "quork." There are "quorks," "quarks," and "queeks."
A call may be short and abbreviated, long and even, or long
and undulating. Calls may be single or strung together in
series of two to six. At any one time, in any one bird, each
series of quorks is of the same kind, and there is usually the
same number in subsequent series.

Aside from the "quork" and its variations and combinations, there are other, more distinct calls. Sometimes there is
a rapid series of percussionlike sounds that remind you of a
piece of metal hitting the spokes of a moving bicycle wheel.
In some birds the percussions have a pure-toned bell- or xylophone-like sound. In others they are more hollow or wooden.
Most of the series of drumlike sounds last about one second,
punctuated at the end with a dull "thunk" or a "pop." Occasionally the sequence is slower, longer than a second, and
not terminated with a thunk. Other variations sound like
rapid lip-smacking or a woodpecker's drumming. It is hard
to believe these sounds are made by birds, most of all by
ravens.

Idiosyncratically, one raven makes a single loud "pop."
Another perches for half an hour in a spruce nearby and
warbles continuously in a softly melodious singsong. What
do all of these sounds mean?

I had heard a variety of raven calls before, but now I
noticed one that I had never heard before except at a kill,
and that one was even more conspicuous here than the drum-

ming or knocking sounds. It is very loud and high-pitched,
like the noise the young make when they are hungry and
calling to be fed. I have no words to describe it adequately,
and I call it simply a "yell." There was a lot of yelling going on.

The feast, though ample now, would be short, and the
time to the next one could be long. It seemed to me the birds
should keep as quiet as possible, so that they would not attract attention and draw in competitors (other ravens, as
well as scavengers like myself) to their rare feast. Large
animal carcasses are not only very rare, they must also be
very difficult to find in these immense woods. How had at
least fifteen ravens found the well-covered up moose in four
days or less? Why weren't they fighting over it until a winner remained, in accordance with the conventional ecological
theory that says animals act for their own individual interests?

I was awed, because I saw a paradox. At this time of year
I had seen ravens flying only singly or in pairs, which meant
they were solitary animals, not likely to have friends or relatives to help, the usual explanation for sharing behavior.
(Ravens fledge in late May, and travel in family groups
throughout the summer.) Furthermore, ravens are not common here, and the chances of even one, by random chance,
flying near any one carcass within several days in minuscule.
It seemed almost inconceivable to me that they (and why
not the much more common crows or blue jays?) could all
have *independently* stumbled upon the same moose carcass.
Could the ravens have *communicated* with each other? If
so, why?

Vultures also accumulate in large numbers at a carcass.
But with vultures, many hundreds of pairs of eyes are involved in the location of any one carcass. Vultures search
from their high vantage points in the sky, riding on rising
columns of warm air thermals. When a vulture in the African plains sees a dead animal, it flies down to feed. Vultures
in the vicinity follow the first and are in turn followed by
others, and so on, until a vortex of vultures converges from
all directions to compete viciously at the carcass. My recol-

lections from watching vultures in Africa were that these
birds go about their business without vocal advertising.

I know that the ravens' situation is different. Although
they soar on occasion, they patrol by flying close to the treetops, and except for pairs and families in the summer, they
are not apt to be in visual contact with one another. A soaring vulture cannot make itself invisible, and thanks to the
many competing scavengers, it must hurry directly to a carcass it finds. But a raven in these woods can remain silent,
furtive, and nearly invisible. It *need* not divulge where it
has found food. Could these ravens, by *not* remaining silent,
possibly be attempting to share their food by calling in the
hungry?

The idea that ravens, who are rare and solitary nesters,
could evolve mutual cooperation in food-finding seemed at
first glance too fantastic to consider seriously. Cooperation
would not have evolved unless it produced a benefit to the
participating individuals. So why should a raven show *others*
where to find food? What does *it* gain thereby?

Given the rarity of moose and other carcasses in these
woods, it seemed to me that if *I* were a solitary scavenger,
I would soon starve. The *only* reason I found the meat was
because others had found it and advertised it. In the best of
all possible worlds, it would be advantageous to all partici-
pants if a large number of individuals searched independently and then periodically came together at a prearranged
place, where whoever had found one of the rare food bonanzas would then lead the others to it. But how could something like that evolve? When I was very young, I was afraid
of the dark. And now this question seemed like the dark,
and it made me nervous.

Evolution is a mechanism for passing on an *individual's*
genes, with those of the species carried along only second-
arily. The interests of the group may be served, just as we
have now "evolved" better cars to our benefit, but the forces
driving the system are not what is best for the group. What
matters is what sells on the marketplace filled with competing models. If only some ravens shared for the benefit of
others, they would be competing against "cheaters" who

reaped the same benefits but did not necessarily give their time and energies. The cold logic of evolutionary theory predicts that the cheaters would reproduce, and their genes would spread at the expense of or in preference to the sharers. A large number of examples of sharing behavior in Nature are based on helping relatives (kin selection) where genes shared with the group blur the distinction between group and self.

Could all of those fifteen or so ravens in front of me be a family? This is very unlikely because ravens have only four to six young, and as far as I knew they disperse in the fall. Something else is going on. The sharers were likely strangers to each other. The incongruity between what I am thinking and what I see looms so large that I know I am onto something big. There is no ready model to accommodate what I see. If I can show that they yell when they find food, and that the others indiscriminately come flocking, it would be something totally new and unexpected. I could surely publish the findings in *Science*, and I can probably get the results in a week or so.

I run through the woods back to camp to bring some remnants of a rug to build a blind where I will be completely invisible to the ravens, who, I fear, might be able to see through the brush. The blind is built in a depression some thirty feet from the carcass and covered with dense brush. And there I spend the whole afternoon, until dusk, on the damp ground. No raven comes near the bait.

OCTOBER 29. I am back again at dawn, but the remains of the moose are gone. Something has dragged them away. Bear tracks! The staying power of meat in these woods is even less than I had thought. Nevertheless, following the drag marks, I find the moose remains some fifty yards up the slope. No ravens are around anymore, and I drag part of the carcass to a different place, about thirty yards west of the original site. I hope the bear will be satisfied with what I have left it. How will the ravens react when they find their "new" meat, after I hide it from them for a day and then uncover it?

OCTOBER 31. I stumble the mile or so through the dark woods before dawn. The first blue jays are screaming, and the evening grosbeaks are already whistling as I reach the bait and pull off the brush, logs, and old green carpet. Then I climb the ridge and settle myself on some moss under a spruce. Will the ravens discovering this "new" food bonanza spread the word?

6:06 A.M. Dawn comes. Flocks of pine siskins fly overhead, whistling in their thin frail voices. A purple finch sings. I hear black-capped chickadees, ruby-crowned kinglets, and a pileated woodpecker. Of the ravens there is no sound or sight.

7:36. Two ravens approach, closely followed by a third. They descend to the meat. No sounds.

7:41. Two more ravens come, and another is close behind. I hear several muted "quorks." Greetings, perhaps, but certainly not advertisements.

7:43. The first three ravens should by now be long satiated. And, indeed, one that has stopped feeding is now "yelling" loudly near the bait. The ravens still congregated at the bait are quiet. The calling bird is perched all by itself, in a tree above the bait. The raven's yelling is interspersed by wild yodeling trills, and the loud commotion continues on and on at earsplitting volume. The yodeling trill reminds me of the trumpeting of a sandhill crane. And the yelling sounds like a dog whose tail is caught in a slamming door. These sounds are unlike any that ravens normally make when they are away from the carcass.

7:48. Another pair of ravens and then a single arrive.

Ravens are not the only birds that come. Now a red-tailed hawk circles; the ravens seem to ignore it, and vice versa. Next comes a goshawk, which takes a more active interest. It chases several of the ravens, which, when it comes near, flip over on their backs in the air. A raven's beak is not to be trifled with, and the hawk does not press its attack. Soon one or two ravens fly up and chase the hawk, which casually flaps off down the valley.

My observations today leave me with few insights. It looks like my paper for *Science* will have to wait.

RAVENS AS HUNTERS
AND SCAVENGERS

I WOULD HAVE LIKED to spend all of my time in these woods watching these ravens. But that would give me a limited and perhaps slanted view of the birds. To get a larger perspective, one must see an animal through the collective eyes of others who have seen it in other places and at other times. And there is only one place to get that broad perspective: in the confines of a research library. To me, this is the most tedious and difficult part of any project.

I could already see that the ravens liked dead moose. But is that unusual for them or typical? What else do they feed from? In the course of this study I read over four hundred publications, looking for information that I thought would ultimately be relevant to my question on ravens. If you have patience, stick with me on these forays into the library. If not, simply skip the chapter and go on to the next, where we will return to the late-autumn woods of Maine.

In *Make Prayers to the Raven*, Richard K. Nelson writes that an old Koyukon man told him: "You know, raven don't hunt anything for himself. He gets his food the lazy way, just watches for whatever he can find already dead. Like in the old story, he always fools everybody so he gets by easy. Only thing raven ever kill is blackfish at a hole in the ice; otherwise I never heard of it kill something for itself."

In the middle of winter a raven in Alaska or Maine may indeed get dead meat the "lazy way," but that need not exclude it getting other food in other ways as well. It would

48

be tedious to try to sum up all the different kinds of food eaten by the raven. Judging from only some of the available sources, a fairly accurate assessment would probably be the following: all the dead animals it can find, all the live ones it can kill, and fruit and grain if they are available. A raven may pick blueberries, follow a pack of wolves and eat their scats, and have at least 285 Mormon cricket eggs in its stomach at a time. Ravens are a common sight at dumps and in some localities even scavenge from fast-food dumpsters in the middle of towns. What is of primary interest is not so much what the raven eats, but the different ways it has of getting a meal. Many birds are evolutionarily programmed to feed on one specific resource in one very specific way. The phoebe, for example, sits on an exposed perch and catches only the insects that fly by. The red-eyed vireo looks for caterpillars under leaves on deciduous trees. The towhee turns over dead leaves on the ground and feeds on the ground-dwelling insects underneath them. The bay-breasted warbler looks for caterpillars at the end of spruce twigs, and the brown creeper searches under loose bark by hopping up tree trunks. The specializations are endless. But the raven distinguishes itself by being a jack-of-all trades, but it can become master of quite a few.

Ravens are quick to exploit new food resources. On Mount Denali (Mount McKinley) in Alaska, they have recently been declared a pest by mountaineers. After reaching 7,000 to 12,000 feet, climbers often bury caches of food in the snow for use on the descent. These caches are usually marked by a green bamboo garden stake tipped with a three-inch-square red flag. Ralph Baldwin, a mountain climber from Alaska, told me of seeing ravens dig down "at least three feet" to retrieve such food wrapped in cardboard boxes and garbage bags. I have no idea how this habit originated. Perhaps initially a shallow cache was exposed by a melt or by a mammalian scavenger who drifted into the lower reaches of the snow fields. After one or two such rewards, the ravens could have made the association between the very specific kind of marker used by mountaineers and food. One association might have been enough. For example, the late great ethologist Konrad Lorenz described how he inadvertently rewarded his pet raven, Roah, with food after it came to him with a piece of laundry from the line. After that, the bird made repeated raids on the neighbors' lines, bringing Lorenz wet undergarments in the hope of getting food. The difference between Lorenz's and the Denali ravens is that the latter were probably rewarded at least a second time.

Like many other corvid birds, the raven may hunt insects and other small prey. Much of its diet has been deduced from the castings of undigested material that this bird, like hawks, owls, shrikes, and some other birds, regurgitates. Such pellet analysis shows that ravens eat almost anything. But although the pellets reveal what the birds eat, they unfortunately do not give much indication of the relative quantities eaten: a raven that has eaten a mouse will void a pellet of the compacted indigestible bones and fur the next day. But if the bird could eat a bull moose, the 70,000 times greater amount of food might still be reflected in no more than a single pellet, because ravens do not eat moose pelts or moose skeletons.

Although pellet remains probably vastly overestimate the extent of the hunting done by ravens as opposed to scavenging from large animal carcasses, they do indicate that ravens eat a great many small rodents, which they probably catch alive. Stanley Temple's analysis of 684 pellets at a raven roost in Umiat, Alaska, indicates that rodents provided an important part of the predatory diet. The pellets also showed remains of caribou (the primary food of wolves) and ptarmigan (the primary food of gyrfalcons), suggesting the ravens relied on scavenging. To my knowledge, there are no published observations of ravens actually catching rodents, but in 1977 Frank Mallory reported seeing a raven apparently attempting to catch meadow voles (*Microtus pennsylvanicus*). And I have seen ravens capture and eat short-tailed shrews (*Blarina brevicauda*) that were tunneling up through the snow to feed at the same meat already being used by the birds. One pair of ravens in Great Britain is thought to have lived "almost exclusively" on moles.

An article by John Johnson in the Los Angeles *Times* of November 14, 1988, described how ravens were apparently decimating the endangered desert tortoise population of the

Mojave Desert by eating the young. Raven populations have grown 328 percent over the past twenty years in southern California, apparently thanks to dumps and other food sources created by increasing human populations.

Ravens may capture considerably larger animals. Stomach contents of eighteen adult and sixty-six nestling ravens near Lake Malheur Reservoir in southeastern Oregon in June 1934 revealed rabbit remains in forty-three of the eighty-four stomachs. Most of the rabbits appeared to have been youngsters less than a third grown, suggesting raven predation. A raven even entered a henhouse in Memramcook, New Brunswick, where it killed "a sickly hen," and another pulled a pigeon from its nest on a cliff. Undoubtedly, confined animals are easy prey for a raven, since even a marsh hawk put into a cage with a raven was disposed of except for "leg bones and the larger feathers" by the following day. A recent report describes a pair of ravens attacking and killing a pair of gulls (kittiwakes) in dim light near Chagnon Bay, Alaska. The famous Dutch ethologist Niko Tinbergen wrote in *Curious Naturalists* that in Greenland "we did see for ourselves . . . they [ravens] chased ptarmigan on the wing and sometimes obviously succeeded in killing them. They also finished the meals of Greenland falcons." He did not state why it was "obvious" that they killed ptarmigan.

Ravens have long been hated by sheepmen for plucking out the eyes of newborn lambs, and Roland Ross reported in 1925 that on Catalina Island in California "the lambing fields were always guarded by armed men." Ravens undoubtedly pluck out the eyes of already dead lambs, but they may also attack live animals. According to Robert Nowsad in his "Final Report on the 1968–1973 Canadian Reindeer Project" of the Canadian Wildlife Service (unpublished):

Ravens normally thought of as scavengers were observed actually killing newborn reindeer fawns. In most cases, the fawn had been abandoned by the cow, or was sick and weak, and was thus exposed to attack. However, ravens were observed working in combination with one another attempting to separate the cow and fawn and thereby making the fawn vulnerable. Varying from year to year, two to three dozen ravens were present during fawning. In 1971, 15 percent of newborn fawn deaths were attributed to ravens. . . . Ravens followed the herd throughout the year, and it was seldom that they were not observed in the vicinity of the reindeer.

There are also other reports of ravens attacking live sheep and reindeer. Ravens may be especially effective as hunters of larger prey when acting in pairs. For example, Chris Maser of the Puget Sound Natural History Museum describes a pair of ravens hunting feral rock doves:

About 100 Rock Doves (*Columbia livia*) have for many years inhabited Succor Creek Canyon, Malheur Co., Oregon. At least one pair of ravens (*Corvus corax*) inhabit the canyon. Over a 2-day period I saw ravens fly toward and dive at groups of doves sitting on projections along cliff faces. The doves were flushed by such maneuvers, but usually stayed close to the cliffs and were not pursued by the ravens. At 19:45 on 14 May 1975, one raven flushed a small group of doves from a ledge, and one broke from the flock and flew 0.4 km across the canyon closely pursued by both ravens. The dove landed on a small ledge near the base of the cliff under a narrow canopy of alder (*Alnus* sp.). Both ravens entered the trees above the dove and flushed it again. They forced the dove into the creek where it landed in an eddy along the shore. The ravens landed on the bank, one upstream and the other downstream of the dove; they chased it back and forth until one killed it with a sharp blow to the head. Without releasing its grip, the raven pulled the bird from the water, and both ravens plucked and ate it. I found several similar piles of dove feathers along the canyon walls and concluded that these piles represented predation by ravens. On 20 May 1975 Sam Shaver and I flushed a pair of ravens from a freshly killed Chukar (*Alectoris graeca*) along Dry Creek, Malheur Co., Oregon. Evidence also suggested predation by ravens.

Similar observations of two ravens working as a team were made by the Wisconsin ornithologist Ludwig Kumlien in 1879 and later used by Arthur Cleveland Bent to "illustrate the ravens' resourcefulness." Bent explains that Kumlien had

. . . on different occasions witnessed them capture a young seal that lay basking in the sun near its hole. The first maneuver of the raven was to sail leisurely over the seal, gradually lowering with each circle, till at last one of them dropped directly *into* the seal's hole, thus cutting off its retreat from the water. Its mate would then attack the seal, and endeavor to drag or drive it as far away from the hole as possible. The attacking raven seemed to strike the seal on the top of the head with its powerful bill, and thus break the tender skull.

Undoubtedly ravens would rather get their meat "the easy way," such as by pirating it from others. The observation of a Koyukon who was watching an otter catch fish is recounted by Richard K. Nelson. When the otter hauled a fish onto the ice, a raven swooped down and chased the otter off its prize. Ravens have also been reported pirating kills from gyrfalcons and owls, and I have routinely seen them give chase and retrieve food from herring gulls, crows, and from one another. Francis Zirrer, an apparent hermit living deep in the Wisconsin woods with a domestic house cat, used to see a pair of ravens near his cabin at dawn, apparently waiting for the cat to return from its nocturnal forays. The birds dove at the cat, which dropped its prey, and then disappeared with their prize.

Hunting live prey and pirating recently killed prey require more finesse and daring than the "lazy way" of feeding on carcasses. The problem is that carcasses are dispersed and rare and not always available in a raven's territory. It seems clear that ravens, especially those in pairs, are able to hunt. But that does not mean that all ravens can get their living by hunting. Complex foraging skills in many birds are acquired only through age and experience, and perhaps in the northern forests, only the older experienced ravens acting as pairs may be able to sustain themselves or supplement their diet by hunting or other foraging skills.

Carnivores provide a large supply of meat, but they eat it fast, too. A pack of six to seven wolves can eat up a deer within hours. But Canadian wolf researcher Robert Stevenson tells me that a moose killed by wolves is often left after about two days, with about one third of the meat still on it. Another Canadian wildlife researcher, Ludwig N. Carbyn, reports that in Riding Mountain National Park, Manitoba, wolves left 40 to 50 percent of available elk and moose carcasses. Other observers also found that wolves leave 40 to 50 percent of the meat; they leave less when killing prey is difficult, and more when the snow is deep and killing is easy.

To utilize the rich resources of meat left after a kill has been made, ravens either follow potential prey and wait until one gets killed, or follow the carnivores until they kill. Ravens are known to follow migrating caribou herds, and Canadian caribou researcher Frank Miller tells me they travel twenty-five to thirty miles per day. Wolves also follow the herds, and since kills are frequent, following massed prey is probably easier and more economical than following the predators as such.

When the prey is scattered, ravens follow the predators directly. The famous wolf researcher, L. David Mech, of the University of Minnesota, who has had much contact with ravens during his classical studies of wild wolves, wrote in his 1970 book, *The Wolf:* "In many regions, including Isle Royale and Minnesota, it is common for ravens to follow wolf packs, wait for them to make a kill, and then feed on it as soon as the wolves leave." When they are following wolves, they "fly ahead of them, land in the trees, await the passing of the wolves, and then repeat the process. During winter a flock of ravens on Isle Royale seems completely dependent on wolves for their food. In Minnesota, most fresh kills are usually covered with two or three dozen ravens unless the wolves are still feeding."

Audrey J. Magoun, studying summer scavenging on the North Slope of eastern Brooks Range in Alaska for her master's thesis, also has strong suggestive evidence that ravens follow wolves. She provided eleven large mammal carcasses, and none of them was visited in less than thirty hours. However, the one carcass she found, "probably less than two hours after the kill was made by wolves," was already attended by ravens. She writes:

The only instance of a raven ever closely approaching me as I was hiking through the study area was the one occasion when I had a dog with me; it resembled a wolf very closely. A raven followed us for nearly a mile, often landing ahead on the tundra and waiting while we caught up with it. It would seem advantageous for ravens to keep close contact with wolves, especially during winter when daylight is so limited in the Arctic.

There seems to be another advantage for ravens in following wolves: Under emergency conditions they also eat their droppings.

Ravens who have learned that wolves are a means to a meal ticket have also learned to draw the right conclusions about them. During a study of wolf vocalizations in the Superior National Forest in Minnesota, Fred Harrington of the Department of Psychology, Mount St. Vincent University, Halifax, Nova Scotia, imitated wolves' howls, and this enabled him to elicit responses from radio-collared wolves. To his surprise, he also elicited responses from ravens:

On 25 October 1972, I found four wolves at the site of a kill of deer (*Odocoileus virginianus*). Between 12 and 20 ravens were perched in the trees directly over the site. At 17:36 I howled and during the wolf's reply many ravens called and flew about within 5 m of their perches. When I howled again 20 min. later, the wolves did not reply, but three ravens flew 200 m toward me in a zig-zag pattern, veering off sharply when 30 m away.

He describes six other incidents where his howling attracted ravens. "In all these cases, ravens (seen flying at a distance) abruptly changed their flying course, approached me, and appeared to search (zig-zag flight, hovering), suggesting that they were attempting to find me."

Harrington indicates that wolves often howl spontaneously at kills and after hunting, whereas summer howling occurs most commonly at home sites. "Although I howled throughout the year, I noted most responses of ravens in fall and winter."

Similarly, pioneer wolf researcher Durward L. Allen wrote

that when Prince Maximilian was near the mouth of the
Missouri in 1833, he observed that both wolves and ravens
would immediately gather at the sound of a gun. A German
soldier stationed in Lapland during World War II also reported that ravens came regularly at the report of a gun
in those areas where there were reindeer. Many Maine deer
hunters claim that ravens are attracted to gunshots. But I
doubt that ravens can tell the difference between a shotgun
blast for grouse and a successful (by far, most are unsuccessful) rifle shot at a deer. In any case, I have never seen
ravens come to a gun's report in Maine, although they sometimes find fresh entrails so quickly that one might easily be
fooled into thinking they do.

L. David Mech also speculates that ravens can track
wolves. It would certainly be advantageous for them if they
could, because they usually stay at a carcass after the wolves
leave, and they do not travel at night, as the wolves do. As
far as I know, however, there is no direct evidence that
ravens follow predators' tracks. It is at least possible that
they can, because in the summer they frequently fly closely
over the human equivalent, that is to say, they fly directly
up and down highways searching for roadkills.

The mystery of how ravens make a living by staying above
the Arctic Circle throughout the polar night of winter presumably also revolves around carnivores. In *Curious Naturalists*, Niko Tinbergen describes being told by Eskimos
in Greenland that ravens followed the polar bears on their
wanderings over the outer pack ice and lived off the remains
of the seals killed by the bears. Ravens also associate with
grizzly bears at their kills in Alaska and in Yellowstone
Park. And, of course, they associate with humans, the ultimate carnivore.

Finally, there is the indirect evidence of the raven's role
as a carcass scavenger through evolutionary history. Unlike
crows and jays, the raven has the long, narrow wings characteristic of many migrating birds; even though it does not
migrate, it appears to be adapted for long-distance travel.
This ability would be extremely useful in finding widely
spaced carcasses. Ravens' large bills (7.5–9.5 centimeters)
are well suited for shearing meat from bones and pecking

into crevices among bones. And their considerable body mass not only increases their competitive ability at the bait against other birds, but it also serves as a buffer against starvation when feeding opportunities are few and far between.

That large body size has evolved in response to carrion feeding is suggested not only in the Corvidae, but also in the Cathartidae (vultures) and the extinct Teratornithidae. The teratorns contained the world's biggest flying birds; the largest known species, *Argentavis magnificens*, weighed 175 pounds and had a wingspread of twenty to twenty-six feet. The remains of over one hundred individuals of *Teratornis merriami* found in asphalt deposits of Rancho La Brea, California, suggest that the birds were trapped in the oil seeps while feeding on the carcasses of similarly trapped animals. But from an analysis of functional morphology of the teratorn skull, it has been suggested that these birds were not scavengers, because they would have been incapable of tearing flesh from carcasses as vultures do. Nevertheless, I speculate that they may have been closely associated with wolves and saber-toothed tigers or with other large carcass openers, just as ravens are today.

I conclude from my reading that even though ravens are adaptable and can do many things, they are nevertheless highly specialized carrion feeders. However, given their intelligence and versatility, their behavior could be intricate and it may not be easy to predict or find out what the mechanisms of their specialization may be.

CALLING IN CARCASS-OPENERS?

> *Then, upon the velvet sinking, I betook*
> *myself to linking*
> *Fancy unto fancy, thinking what this*
> *ominous bird of yore . . .*
> *Meant in croaking "Nevermore."*

—EDGAR ALLAN POE,
The Raven

THROUGHOUT THE FIRST year of my research, and especially during the following years when I was no longer on sabbatical leave, I had to return to my teaching job at the University of Vermont, to my family, and to the library, and I declared several temporary truces between the ravens and myself. The first time I returned to Maine after an absence of ten days, I was still full of questions.

NOVEMBER 10, 1984. What can the yelling mean? It must have a function. How can intelligent birds ignore it? If only I could *prove* that the ravens' yelling indeed attracts others! Ideally, one should record the ravens' sounds at food, and then play the recording back in the absence of food to see if it attracts ravens from miles around. The problem is, it might be days, or weeks, before a raven is again "miles around" to hear any such call.

This time I come equipped with a borrowed cassette tape

recorder, directional microphone, parabolic reflector, and loudspeaker that can play back an impressive 70 decibels at 250 feet. All I need now is another moose. But surely ravens aren't fussy. A friend has just slaughtered a pig, and the donated entrails ought to do just fine. Forty pounds of pig guts should be a big enough bonanza to move any raven to become generous. After one raven finds them, I will record its yells to play back later. It should be simple. But how long might I have to remain rooted in a blind before a raven comes? A day? A week? Maybe I am foolish to work on ravens. "Ravens," said the mentor of a professor friend of mine, then a fledgling graduate student and raven enthusiast, "are smarter than you are, and it will take you years to outwit them enough so that you can begin to get meaningful data." There was more than jest in these remarks, and my friend distinguished himself in his profession by figuring out the ways of caterpillars, instead.

Would I have the fortitude to lie on my belly in a damp blind to watch those pig guts for days on end? I doubt it. But possibly ravens would still be patrolling in the area of the moose remains. These thoughts induce me to pack the entrails into my knapsack and tote them through the mile of woods, up the steep ridge, and down again to near the moose kill.

There is not much left of that moose. Every scrap of meat has been picked off the ribs and hide. However, the birds have not penetrated to the remaining pounds of meat in the head, which have not been skinned or torn by a carnivore. Only the exposed eye is picked out.

I spread out the entrails on the ground and return to my old perch under the spruces with my tape recorder. The gray dawn passes. The moss hummocks again turn from black to luminescent green under the hazy overcast sky. Fog drifts in. The hours go on and on.

Near noon, two large black birds appear silently as if out of nowhere. They shoot in among the birch trunks and descend near the meat. They croak very softly a few times, almost in a whisper. Then all is totally silent again. Minutes. A half hour. The sudden "swoosh, swoosh, swoosh" of strong wingbeats rends the air. The two ravens fly up, croak a few

times, then meet up with a third. All three career up and
down the slopes, rising swiftly in the air, then plummeting
down with tucked-in wings. In a minute or so they are a
mile away. What was this all about? Have they left to get
their friends?

More hours pass, and now I am shivering. Why did no
new ravens arrive? Have they found another kill? I repack
the entrails and return to camp to use my partially built
cabin as a blind.

NOVEMBER 11. At dawn I am already kneeling and facing
east, peering through a crack between the logs. My eyes are
focusing on the pig guts seventy-five yards in front of the
cabin at the end of a clearing. The hours drag. I see no
ravens and hear no ravens. Near noon I leave to go to Camp
Kaflunk to heat up a can of beans and a cup of coffee.

I don't know if it was the raven's perversity, cunning, or
my bad luck, but after I left and had just built my fire, a
raven's call sounded from the bait. Lesson number one: You
can't leave for a minute or you don't get to find out what
really happened. Returning swiftly through the woods, I
am in time to see a raven perched on the top of a fir tree
near the bait. It is still making an odd assortment of bizarre
noises, but before it departs, I manage to record some of its
fantastic sounds.

Later that afternoon I see three ravens in the area. They
fly over the clearing and the bait and perch in the trees
nearby, but none of them comes down to feed. Nor do they
yell. Why are they quiet now?

Aren't they hungry? This is the middle of the hunting
season. Fourteen bear and eight white-tailed bucks have already been tagged at the local game tagging station. The
entrails of these animals, plus those of the wounded that died
later and were not officially recorded, are probably scattered
around in these woods.

NOVEMBER 12. The day starts off badly. I hear a raven's
croak before I am even out of my shack at 6:30 A.M. Perhaps
it was already at the bait and won't be back for the rest of

the day. That will make the waiting all the harder. Lesson number two: Get up *early* to start watching.

At 6:45 A.M. I am lying down beside my tape recorder, again looking out through the unchinked logs. But I hear only blue jays. At 8:15 two crows arrive. After swooping over the bait, they land in the trees and silently appraise the situation for thirty-three minutes. Then one, perhaps braver than the other, descends to the bait, walks directly up to it, and tears off hunks of fat. The other watches from its perch. The crow, with white fat protruding from its bill, flies off into the woods, presumably to cache this food. The other soon joins in, and from then on for the next two and a half hours the two work steadily, making a total of forty-four separate caching trips. They take no meat, only white fat. They are silent the whole time, possibly setting a record for crow reticence. Apparently *these* highly social birds have no intention of sharing as they quietly sneak off with their loot.

The ravens, meanwhile, have not forgotten the bait. At 8:55 A.M. one comes by, silent except for the loud swooshing of its wingbeats as it flies low over the clearing. Its bill is held down like a skimmer's, and its head darts in rapid scanning motions to the right and the left. I am impressed by the glistening sheen of its feathers; they look like polished metal. But the bird does not stay around very long to be admired. It dives at one of the crows on the bait and chases it down the valley.

A raven returns at 9:14 A.M., and then again at 10:37. Each time it simply flies over the clearing, as if rechecking the bait. I suspect that it is the same bird, because it comes a little bit closer at each return. At 11:12 it finally flies down to the ground near the bait, while a second one, which has come along this time, perches close by in a maple tree.

The raven near the bait hops up and down in a few nervous vertical leaps before tearing off a chunk of meat. Without bothering to eat, it flies off into the woods with a fist-sized portion. In a few minutes it returns. In the next fifty-three minutes it and its companion make a total of eight more trips, carrying off other portions. Like the two crows, both ravens remain *totally silent* while feeding and/or making caches!

These ravens are disappointingly not performing accord-ing to my hypothesis: They showed no recruitment. Should I get up and leave, admitting my idea is wrong, and forget about publishing in *Science?* There is no quicker way to scientific oblivion than to cling to ideas in the face of con-trary evidence. I've just seen contrary evidence.

One or two ravens come back several times in the after-noon, as if checking to see if the bait is still there. But they do not descend to feed, to cache more meat, or to call. There are usually two birds, and they quork occasionally, but only while in flight, as if keeping in voice contact with each other. Will they begin to share with neighbors tomorrow? Maybe they call in others only after they have salted away a surplus for themselves or a mate.

NOVEMBER 13. On this the fourth morning I arrive a little earlier, at 6:20 A.M., hoping to beat the ravens. I succeed. At 6:44 two of them arrive flying in close formation, trailed by a third. From my cramped quarters I cannot see them at all times, but within the next half hour I twice hear a low soft quork. One or all three of them are perched nearby. At 7:24 two ravens perch on an old red maple tree above the bait. One preens for twenty minutes, while the other keeps looking in all directions. Then at 7:33 all three leave, flying back down the valley. As before, it seems they came neither to feed, nor to cache, nor to call in others.

I now have yet another hypothesis: Maybe the ravens' yelling is not a call to share the feast with other ravens, but to recruit other animal scavengers (i.e., coyotes, bears) to open the carcass! And other ravens merely get in on the message. The pig guts lay exposed. There was no need to call in help to tear into the meat.

Within a few days or weeks the first snow will fall, and it usually accumulates to more than six feet. Many deer will be crowded into yards where the forage will become depleted. The weak, the old, and the diseased will die of starvation. These animals, as well as predator kills, will be the ravens' link for survival through the winter.

The carnivores do not live off only the kills they make themselves. They also scavenge, eating the weakened and the dead, which are well preserved in the winter. The ravens, through their aerial surveillance, could perhaps locate dead animals for the carnivores and call in these living hide openers. The ravens, in turn, would be repaid ecologically by feeding on kills brought down by the carnivores.

I later learned that Frank C. Craighead, Jr., in his 1979 book, *Track of the Grizzly*, had the same idea during his studies of the Yellowstone grizzly bears. He wrote:

> We often noted the presence at a carcass of ravens and magpies which fed when the bears' appetites were satisfied, and quite regularly the ravens were on the scene even before it was discovered by the grizzlies. They would circle above it, then land, and the first to approach would consume the eyes, which were easily available. After this it was nearly impossible for the ravens to get more sustenance until the carcass was torn open. At one elk carcass within Marian's home range, the first raven to spot it took off, and before long there was a growing assemblage of the large black birds. I am inclined to believe that in some way they announce the presence of a carcass to both coyotes and bears.

A message by a raven meant for carnivores could easily be exploited by other ravens, leading to (inadvertent) recruitment. If this idea is correct, then ravens should call loudly at large carcasses with *inaccessible* (i.e., "unopened") meat and keep quiet when the food is available for easy picking, as it is at these pig guts.

I would be surprised if coyotes and other predators did not respond to sounds that are reliably and *specifically* associated with food, like raven's yelling. If not, Pavlov would turn in his grave.

There is a famous precedent in the African honeyguide, a little bird that specializes in feeding on honeycomb and bee grubs. The honeybees nest in hollow trees (and sometimes in the ground) where the weak-billed bird, although related to woodpeckers, cannot penetrate. Having found a bee nest, the honeyguide finds a honey badger or a human bee hunter and by a loud chattering call leads these predators to the bee tree. The predator tears open the bee nest, and the little bird later feasts on the remains.

My four days with the pig guts have not been a rousing success. Instead of an answer, I now have another hypothesis. I cannot wait to do the next obvious experiment: to provide the ravens with a fresh *whole* carcass. If my new hypothesis is correct, the raven who finds this carcass will yell interminably until a coyote or bear comes. Other ravens will come as well.

NOVEMBER 21. I have been away in Vermont for eight days to recoup and to buy a goat. The goat cost me $60, a very low price indeed when you are about to make an important discovery. This Alpine Toggenburg goat is as big as an average Maine deer (150 pounds), and with its brown back, white belly, and short white tail it looks very much like a doe as well. "Good enough to tag," they would say. People stare when I drive out to Maine and stop at a diner with a goat strapped on the hood in hunting season.

It is dark before I finally manage to drag the goat up to the clearing in front of my blind. A bear, meanwhile, has removed the remnants of the pig guts, leaving deep pug marks in the softened soil of a large flattened mound.

I hardly sleep in eager anticipation of the dawn, imagining over and over a raven majestically coming in from the forest, landing by the goat, yelling loudly, soon to be joined by others. The scene plays again and again in my mind— and then I am suddenly jolted awake by the thought that a bear could remove the carcass before morning. I can take *no* chance of having a bear spoil tomorrow's fun, and I tumble out into the dark and drag the goat a quarter-mile and into the shack with me.

NOVEMBER 22. It is not only the goat's smell that rouses me before daybreak. It is also the smell of victory. Everything I can think of predicts that if I wait long enough a raven will find the carcass, and that raven will yell. But predictions are cheap. Observations are dear at 5:30 A.M. at subfreezing temperatures.

Temperatures dipped to 15°F last night in my shack. The goat is stiff and even more difficult to drag back the same quarter of a mile it had come just a few hours ago. This task is especially difficult on an empty stomach and without coffee. It is now a race with time because I can already see a faint glow of dawn on the eastern horizon. I cannot afford, under any circumstances, to have a raven see me today. Because if one does, and it doesn't yell, I will have an excuse for it.

The eastern sky is orange now. The white frost glistens. My feet are already aching from the cold. It is not yet 7:00 A.M. Low over the fir trees from the direction of Hills Pond comes a raven with fluid, powerful wingbeats. It lands in a tree next to the goat. One minute later, it flies down, lands within five feet of the $60 goat, walks up to it, gives it a quick peck, and flies off—as totally silent as it had come.

At that moment a good many of my hopes and expectations leave with it. Now I am totally befuddled. I *know* ravens yell at food. Surely the calls are not *meant* for others who might be competitors for food! That would be preposterous. The more plausible hypothesis looked so good, so totally consistent with what could be predicted. Now I do not know what to think or what to test. Well, there is always room for individual variation.

NOVEMBER 28. I'm back, giving it another go with the goat, after it has spent a few days in the cabin. Nothing all morning. At 1:10 P.M. two ravens fly over, silently. Then in the distance I hear "quork, quork, quork, quork," the presumed contact calls. The other bird answers with only two quorks, but lower pitched. I am beginning to recognize these very distinctive calls; I think they are the same individuals who were here weeks before at the pig guts. Only one thing is becoming clear: The unopened goat is of little interest to these birds.

Could my favorite hypothesis be correct but the results negative simply because the birds are not hungry? Perhaps their behavior is the human equivalent of saying: "Why should I bother to yell in front of this old goat, when I have food caches immediately available and know of gut piles all through these woods?"

Now I know I must try it one more time *after* the hunting season, using the same goat. It will keep well in the deep-freeze, my shack. In the meantime, I'll go back to Vermont and look at crows for a while. Sometimes one can learn a lot about the selective pressures that shape evolution by studying closely related species, each fitted for a different niche. It is much like physiological experiments where you observe the effects of different variables.

The local blue jays and American crows have their own, unique foraging adaptations so they can make it through the winter. With the snow a few inches deep already, blue jays are still digging down to retrieve acorns they had hidden weeks before. At a local crow roost I watch thousands of birds leave at dawn and come back in the evening. Opening up regurgitated pellets from under the roost, I recover seeds, showing that in mid-December the birds were foraging primarily on the white berries of panicled dogwood and wild grape and (to my surprise) on sumac. In all of the 260 pellets that I analyzed I found not one hair, feather, bone, or scrap of insect cuticle. It seems the crows switch to a largely vegetarian diet in the winter. By contrast, the two raven pellets I later found under a raven roost in Maine contained primarily hair and bones. But ravens do not eat moose and deer bones, the way they do the much smaller bones of mice.

Not adapted to live off large animal carcasses in the winter, neither crows nor jays show the same behavior at the carcasses that ravens do. Neither recruits. The jays are downright belligerent. One (or a pair) does not tolerate others at any food pile it finds. (Although at permanent food stations a number of birds may gradually get to tolerate one another.) If ravens recruit, this behavior is something special, not just a corvid family trait.

DECEMBER 20. It is five days before Christmas, and much has changed since I was here a few weeks ago. Hills Pond is covered under a thick lid of ice, and this and all the rest of the woods are under a deep blanket of snow. The deer will not be weakened by starvation yet, so this may be a lean time for both the predators and the scavengers. Will it also be the time that the ravens are the most willing to share their finds by recruiting? Or will they share less? I predict they will *now* recruit, if food sharing is indeed an evolved behavior that functions to promote survival on scarce, scattered bonanzas.

I am hoping again to get a clue about when and why the ravens yell. I still have the goat from the month before. I have also brought along two more goats. These are not yet frozen solid, but they will be soon enough, unless I sleep with them in my down sleeping bag, and I shudder at the prospect. I therefore start my experiments quickly, because I do not yet know whether or not ravens eat frozen meat.

The brown Toggenburg looks no different after spending the month in my shack, and I drag it out into the middle of the clearing. It should be an easy target for any passing raven. I also put out a few scraps of meat fifty paces away from the goat to find out what the ravens might prefer.

One day of waiting: nothing.

On the second day a pair of ravens comes. Excitement at last. They quickly eat the scraps of meat, paying no visible attention to the goat! Nor do they show any interest in the next two days. I again conclude that they have no use for an unopened carcass. But because they don't yell for help, this should also be sufficient to demolish my hypothesis that their yelling is meant to call in carnivores to open up or dig out carcasses. O.K.—one loose end is tied up. But then another thought intrudes: Can it be possible that these birds want moose or deer and learned the last time during their close reconnaissance that my goat was a poor imitation?

The question is too important to be left unanswered. I must present the goat to *other* presumably naive ravens at a different site to see if I get the same results. I drag the goat the half mile back to the road, put it in my jeep, and then drag it into another field along a lone country road a few miles distant. Having found out earlier that ravens do not come near a parked vehicle (although they seem to ignore *moving* traffic), I hide myself in the dense branches near the top of a tall white pine along the road to begin another watch.

As usual, after a few hours I am anxious for something to break the monotony. It comes. A car with three men passes, slows down, and backs up. A man looks out the window at the goat in the field. Another one gets out and runs across the field toward the carcass. When he gets close, he hollers back, "A big deer—a doe." I am elated. If people are fooled, then surely a raven would not be more discriminating in the carrion it would eat. The man in the car, thinking of poachers, wants to know if she has a bullet hole. The other rolls her over. "Frozen stiff," he declares. "Can't find a bullet hole." Then he must have seen the udder. "It's a *goat!*" At that point I can no longer hold my breath, and I break out in a chuckle. Looking up at me in the tree, the man yells, "What in hell are *you* doing?" "Watching ravens." As he jumps back into the car and is about to slam the door behind him, I hear, "Some pervert up in a tree watching ravens. . . ." I settle back into the cold wind among the branches, a happy "pervert" warmed by another tiny clue to help decipher the puzzle.

It seems almost certain that the yelling is not for scavenger-recruitment. I can try to disprove the idea even more decisively later, but in the meantime I'd better go on to other things that are more likely to lead to the *real* answers.

Where to begin? Nothing so far makes much sense, but almost anything might be relevant. Any one aspect of feeding behavior is apt to be tied in with all the rest of the biology of the animal in a coherent package. Somewhere there must be a thread running through it all. That thread is not yet visible, and all the unexplained loose ends remind me of an untidy room where bits of treasure are mixed in with a lot of trash. The treasure, if assembled, is a delicately running watch, but the pieces are lying every which way, and I don't know how to put them together. Worse, I cannot yet distinguish between the trash and the parts of the watch. I have to "see" how the watch runs before I can be selective in the pieces I pick up and try to piece together. My ability to spend all this time watching is in direct proportion to my excitement over the problem and my knowing that there *is* a "watch" to be found. Every detail, every nuance of what I observe may assume significance because I'm constantly fitting and refitting what I see with what I expect to see.

A colleague whom I told about my methods of studying ravens immediately told me I should apply for a grant to get money to automate my observations, perhaps by installing video-cameras to substitute for my own presence. Undoubtedly, getting the trappings of science would make the work seem more "scientific," but I know it would make me a slave to the equipment; it would also make me lose sight of whole dimensions of behavior that the equipment is not designed to see. It sees only what you design it for or set it up to record. There is still no substitute for direct observations, especially when you don't know what the relevant variables are.

A SELFISH HERD?

Thus I sat engaged in guessing, but no
syllable expressing
To the fowl whose fiery eyes now burned
into my bosom's core. . . .

—Edgar Allan Poe,
The Raven

WHILE I WAS BACK in Vermont, I played my raven
yells through a speaker at an open window to test the equipment. Was the speaker loud enough? Half a mile down the
road I could still hear it. The yells carried quite well, and
I thought it would be fun to try it out in Maine. But I found
the experiment was already in progress locally!

As I was walking back to the house, I looked up. A raven
was flying directly over the house, turning its head this way
and that! I had never before seen a raven near our house in
Vermont. Was the bird there because of the tape I played
for several minutes?

The raven flew on, but as I entered the house, I heard the
familiar "quork, quork, quork, quork." It was answered by
a second one farther away, with a deeper "quork, quork."
Apparently the first raven had only made a reconnaissance.
It now came back with a second one. When one of them
landed in the sugar maple directly beside the house, I realized that these birds' appearance was *not* just a coincidence.

The Maine raven's recorded calls had attracted a stranger. I was euphoric. My initial hunch must have been right. Birds at a valuable food resource *called* in others, who could be strangers! But you can't base conclusions on interesting anecdotes, no matter how compelling they seem.

JANUARY AND FEBRUARY, 1985. It is getting more and more fascinating all the time, and I am itching to get conclusive results soon, so I can take a rest. It is a race now, and the finish is in sight. Maybe I heard no yelling at the baits before because my bonanzas were too small?

The kerosene lamp illuminates the cabin with a dim glow, as I write up my notes after another long winter day. It has been a cold day and not a very eventful one. In several hours I will face yet another icy dawn like many others to await ravens.

Earlier in January I had set out road-killed cottontail rabbits at two different locations. My obvious question was whether or not yelling and recruitment are related to the amount of food that can be shared. The rabbits' meager body mass seemed rather skimpy fare for a raven to want to share.

As I had anticipated, the two ravens that eventually found the two rabbits behaved almost alike: They came down to feed several hours after first flying over, and they remained silent near their prize. These were not very exciting results because the lack of recruitment could have been merely lack of time. The two ravens did not leave any rabbit after the first day. And one day may not be enough time to see recruitment. So little, if anything, was really tested.

EARLY MARCH. *Really* to find out if recruitment depends on the size of the food offered, I may have to provide a super-bonanza. Thanks to a trapper, the local game warden, and a farmer, I have assembled three skinned 30- to 50-pound beaver carcasses, a 175-pound white-tailed buck, and one 50-pound goat. I'm again convinced that doing research is mostly hard labor, and with fieldwork it's hard *physical* labor, usually without pay. It's all for fun, right? In any case, for the moment my fun consists of struggling to drag carcasses a half mile through deep snow to the ledges that I can watch from my shack. But as usual, it is much more difficult to sit an indeterminate number of days waiting for things to happen. One thinks time is life, and you wonder if you really have all the time in the world.

MARCH 9, 1985. Not much happened. Only a pair of crows stopped by briefly.

MARCH 10. I awake in the night to a sharp cold north wind, as heavy sleet is turning to snow. By daylight the snow is falling so thickly that I can scarcely make out the lumps of covered carcasses down below on the ledges. I periodically brush the snow off. Hours pass. Perhaps a raven will find the carcasses today and yell.

There it is—a huge black bird! It dives down through the branches, lands beside the numerous carcasses, and hops up and down and backward in the dance I've by now seen ravens do every time they find a carcass and before one of them begins to feed. In a few minutes the bird is tearing and gulping. Then it flies off with pieces of fat dangling from its beak. It makes five caching trips, each about ten minutes apart.

The raven is *totally* silent the whole time! This is not what I had expected, predicted, and hoped for. Is it possible the snow prevents this bird from realizing how huge the food bonanza of five carcasses really is? Was the noise at the moose and other large meat piles an aberration? No. *Every* observation is significant. Each has meaning.

I can't avoid the obvious: There is no yelling and no recruitment by the discoverer. Maybe my recruitment hypothesis is now finally and thoroughly disproven. I'm depressed. There is nothing to do but pack up and drive the two hundred miles home, back to comfort and civilization. But just to be convinced that there is no recruitment, I this time leave all the meat where it is.

During the long drive on icy, snowbound roads, the nagging suspicion gnaws at me that this is the end of my experiments. Have I wasted a lot of time, or is it still possible that the discoverer would fully satiate himself or herself, and *then* recruit?

Four days later my graduate student Steve Smith kindly offers to drive to Maine to check up on the meat pile. To my utter amazement he comes back and reports finding only stripped skeletons! When he first got there, a dozen or more ravens were still perched in the trees, and others were yelling loudly in the vicinity. But obviously many more must have been there. A dozen ravens cannot eat several hundred pounds of meat in four days; recruitment probably *had* occurred after all.

I am clearly missing a big piece of the picture. Until now I had thought that the yelling attracted birds who just happened to be flying nearby. Anyone who has ever spent any time at all in these woods knows, however, that ravens are seldom "nearby." Something else must be going on. Maybe the birds recruit from far greater distances than the mere range of their voices. If a bird really wanted to recruit, it could do no better than to collect followers at a nocturnal communal roost. Do ravens here have communal roosts?

One of several possible functions of communal bird roosts is that they serve as "information centers" where unsuccessful foragers follow successful ones. But in the many studies on birds made so far, all information-transfer that has been demonstrated has involved the pirating of information. The birds presumably don't sit around and say where they will fly the next day. Information may be given unwittingly. For example, successful ravens could exhibit soiled breast feathers, full crops, or some behavioral cue that would induce unsuccessful foragers to follow them out the next morning. Alternatively, they could betray their knowledge by their behavior, such as showing excitement or eagerness in leaving in the early dawn.

Giving information on food location could at first be incidental, with the signals becoming more specific and the reading of the message more precise so that real communication evolves, *provided* it is mutually beneficial. Such a communication system has evolved in honeybees from mere intention movements to leave the hive. These have become ever more precise and stereotyped until they turned into the "dance," a *symbolic* flight giving information of the actual flight to the food. The hive as an information center helps the bees exploit highly temporary food bonanzas. Perhaps the ravens' yelling marks the precise location of a carcass much as the scent marker (pheromone) that bees leave identifies the precise location of a flowering tree. Yelling at the time of food discovery would be a waste of time and effort if the potential followers would not be contacted until that night, some miles distant. Might *that* explain why a raven that discovered the food did not yell? If the honeybee analogy is correct, then ravens yell only when they already know that their *relatives* are close enough to be heard. But how can solitary nesters like ravens who nest miles apart have relatives close by?

MARCH 30. The winter is almost over. This time when I leave the ravens a superlarge food bonanza, I'll stay long enough to watch the process of recruitment. It will be my last chance this year. In a short while the snow will melt, spring will begin, and the ravens will switch over to a diet of insects, frogs, and other small prey, so no recruitment can then be expected.

My offering this time is two cut-open Holstein calves. Together they weigh nearly four hundred pounds. I place one calf seventy yards north of the shack, the other just three yards beyond it. (It turned out that their relative positions gave useful information.)

Early in the first dawn a group of eight crows appears, cawing loudly. Undoubtedly they are a group of returning migrants. They briefly investigate the calves and fly on. At 8:10 A.M. a raven flies over and calls briefly. An hour and a half later a raven flies by more closely, calling more frequently and yelling at intervals of a few minutes. Three hours later the yelling resumes, and two or more ravens continue to be present on and off throughout the afternoon. In late afternoon I see four of them fly over simultaneously. There may be more in the woods nearby. The yelling continues. Then it is silent. Suddenly the silence is broken by a high-pitched knocking sound like a metallic drum. Is it a lookout giving the "all-clear" signal? The drumming continues at intervals, and I see a raven (but not the "drummer") approaching stealthily on foot over the snow. Three times it gets almost up to one of the calves, hops up and down, and flies up nervously, only to return in a few seconds. One of the other ravens continues to make the knocking sounds. But no raven touches the calves.

A pair of crows who show less caution comes, making seventeen visits to feed and to tear off meat to cache. Almost invariably, one member of the pair calls loudly from a tree, while the other feeds silently at the carcass closest to the cabin. The farther carcass was not once visited by any birds. What happened to the eight crows?

MARCH 31. Again, a group of eight crows comes at about 6 A.M. But this time they stay all day long. They feed exclusively at the closer carcass until 8:30. They go to the carcass as a group, and they get along amicably. Later on in the day they sometimes feed one at a time or in smaller groups, and they now visit both carcasses equally.

A raven was already calling from the woods at dawn, at 5:15, and I heard others all morning long. One swoops over the bait at 10:10. At 11:30 I see five, and at 11:50 there are at least six. At 11:58 four ravens descend and walk cautiously side by side toward the closer carcass. Crouching close to each other, they advance hesitatingly with outstretched necks, until one pecks the calf. Instantly they *all* jump up and take flight. Within fifteen seconds they regroup for the same maneuver, and this time five more swoop in and join them. There are even more about. I hear the knocking calls continuing to ring from the forest nearby. They sound like jungle drums, and I feel that the atmosphere is tense. The ravens fear the bait, but the crows have been feeding all morning and seem by now totally unconcerned.

No raven wants to be first at the bait. One, however, eventually hops up and down, doing a "jumping-jacks" rendition on the snow next to the carcass. An animal evolved to specialize on carcasses for which it hunts visually from the air might be expected to be attracted to dead and sleeping bodies alike. But it should then have a means to differentiate the living from the dead. Perhaps the jumping-jack maneuvers express a raven's understandable unease in approaching an unknown furry creature, but there is no reason why the unease needs to be expressed in such an obvious way, unless it serves a function. That function could be in drawing a reaction from a live animal, thereby letting the approaching bird know it is perhaps not safe to proceed any farther. Or a direct approach (whether with caution or not) in a carcass specialist who is itself palatable could be exploited by the evolution of death-feigning as a hunting strategy in a carnivore.

Soon the jumper is joined by a half dozen others. Nine suddenly begin to advance together. Just as suddenly they take flight as a group. Within several seconds they are back. The first one hesitates less and less; soon there is a group of ravens at the bait almost continuously. Feeding finally begins at 12:05 P.M. At 1:45 the feeding frenzy is over. I see twelve ravens simultaneously in flight, and the birds disperse in different directions.

These observations support the "selfish herd" hypothesis first advanced by British zoologist W. D. Hamilton at the University of Michigan in 1971, as an explanation of why animals gather in crowds. If there is a given amount of risk, then the larger the crowd you are in, the more the risk is diluted. Perhaps the ravens feared something at or near the bait, such as a lurking predator, and they recruited in order to get enough members to minimize their individual risk while feeding. But what can such big powerful birds fear? Blue jays and crows should have much *greater* reason to form selfish herds at food. Also, the hypothesis predicts that recruitment should be *before* feeding begins. Once the bait has been found to be safe, the recruitment should stop. We'll see.

I had wondered before about the mechanism of recruitment: Do naive foragers follow successful ravens from a nocturnal roost to a carcass because they have examined their roost mates, smelled their breath, observed their full crops, or noted their soiled feathers? Now I know the answer is none of the above. It seems that substantial recruitment occurred long before *anyone* had fed! None of the recruiters could have had telltale marks of the anticipated meal upon them. So it was their *behavior* that drew the others here.

The difference in behavior between crows and ravens in this fortuitous experiment is instructive. Since both groups of birds were here at the same time, in the same area, at the same bait, it was an ideal "controlled" natural experiment of the kind one could seldom duplicate if one tried. For the whole day the ravens never so much as perched at the second nearly identical carcass that was only three yards beyond the closer one they eventually fed from. They often landed near it but took pains to walk *around* it when they came close in order to get to the first—and as it seemed to me, identical—carcass. They had taken a day and a half to decide *one* was safe, and they stayed with that one only. That the other was of exactly the same kind made no difference.

Unlike the crows, the ravens seemed to have a need for company before they dared to feed at this, to them, strange bait. Not one raven fed by itself, at least not on this day. But why did the bird I saw two weeks ago fly boldly in by *itself?* Were the birds more afraid because they were not familiar with Holstein calves?

The crows and ravens at no time fed together, but they often perched close to each other in the trees near the bait. When the ravens were at the closer calf, the crows fed on the second calf. If a raven walked near the second calf, the crows jumped off without being chased. The ravens were clearly dominant over the crows, but they made no overt attempt to disturb them.

There was another important difference. The bait was discovered by one raven, and raven numbers increased gradually throughout a day and a half until the birds were confident enough to go in and feed. But when one pair of crows found the bait, only that pair stayed and fed, and when there were eight crows, it was because the eight came originally as a group. Is the lack of recruitment in crows somehow related to the fact that crows are frugivores who only incidentally feed on carrion when other food is not available?

A bird that feeds on berries and invertebrates does not need to be aware of its "prey" attacking it. But one that feeds on large animals has to make sure that what it pecks is not still alive and kicking, attended by lurking predators, or booby-trapped. Maybe their food sharing was a form of trading where a successful forager in effect tells a hungrier raven: "I'll show you where there might be something good to eat, but only because I expect you'll be willing to take the risk of going in to see if it is all right." It seemed to me significant that yelling to recruit occurred *before* feeding began. But then again, I now also knew that much yelling (and recruitment) occurred *after* feeding had begun, when the ravens no longer showed signs of fearing the bait.

Evolution of a function for one purpose does not preclude that function from being changed, modified, and expropriated for another one. Birds' feathers, for example, may have originally evolved from scales as a sun shield or as insulation from either heat or cold. It was then possible for them to be further modified as structures for flight, for tools in sexual and other signaling, and—in sand grouse and ravens—for transporting water to the young. Similarly, the ravens' recruitment behavior for originally "selfish" reasons could also have resulted in mutual sharing of rare bonanza-like resources. It is conceivable that ravens' food sharing *now* might have a dual function.

It seems that the more I see and the more I think, the muddier the water becomes. I had hoped for the opposite.

APRIL 1. It snowed at night. Five inches had accumulated by morning, and at dawn six ravens and the eight crows had already arrived. Pairs of the ravens are perching on the snowy spruce branches near the bait, billing each other and making soft comfort sounds. Crows and ravens fly repeatedly over the now-buried baits.

One raven lands by the nearer carcass, jumps up and down nervously, and leaves after making a few swipes of his closed beak back and forth, shoveling snow. By 6:00 A.M. all of the ravens have left. One or two occasionally come by later throughout the day, as if to check on whether the meat has again become available.

I am tempted to uncover the carcasses, but this is a natural situation to exploit for experiments. The ravens seem to be motivated by a search for food, or else they would not be here. That is a condition one could never be sure of in the field. Now I can see: Will they dig out their food? Will they yell, possibly to recruit coyotes to do the digging for them? Can I call *them* in with playbacks of the yell? Answers to the first two questions are negative. The answer to the last is positive. I'm reinvigorated.

On six different occasions when I heard a raven in the forest, I played my tape of raven yells. After four of these broadcasts, one or two ravens appeared within fifteen seconds, even though no ravens were seen for a timed fifteen-minute "control" period prior to the test. Here is more confirmatory evidence that the yells given at the bait serve to attract others. It makes the unanswered questions all the more exciting.

The birds' fear of baits and their recruitment are both very bizarre behaviors. Somehow they are likely to be inter-related. The experiment this time is in accord with the selfish herd hypothesis. But I'm not yet convinced of this as the only explanation, because if this is it, then yelling, and recruitment, should stop quickly after feeding begins. This is not the case. Something very weird and exciting is going on here.

A CORVID COMPARISON

OCTOBER 31, 1985. THURSDAY. The deer hunting season starts this weekend in Maine. The ravens will not yet have been sated by hunters' leavings, and it is a good time to resume my studies for another winter. I eagerly start for Kaflunk.

Stopping halfway in St. Johnsbury, Vermont, at Anthony's Diner, for the by-now traditional cup of coffee and a woodsman's burger, I feel full of excitement again. If anyone asked me what I was doing and why, I could give only a lame answer. To me, the paradox is stark and beautiful, like the ravens themselves, a real thing to be admired: The ravens loudly call "food," where, by all the evolutionary selfish reasons that I expect them to have evolved by, they should be silent. Why do they advertise? There are many very different, conflicting observations already, but I am sure they are all interconnected to make one coherent story or mechanism of survival. I doubt that I can really understand any of it until I can visualize all of it.

Right now I still have nine hypotheses. Not theories, just hypotheses. It could take me a lifetime trying to go through this list of hypotheses exhaustively testing each one, one at a time. And I might *still* not have the answer at the end. I have to go on hunches. But I have to work to get the right ones.

Every trip is exciting because I try to set up an experiment to "test" one of the different ideas. But they are not rigorous tests—just probes to get at the general features, to allow me to have good hunches so I can later bear down on the *relevant* ideas. Most of the ideas will turn out to be false, so there is no need to mention them formally and in detail because I will reject one after another anyway. Gradually the focus will get sharper as I pare away the chaff. Only the answer is of interest, when all is said and done.

Ironically, one focus on the problem will require seeking out and documenting what some other related birds, such as crows and blue jays, do. The comparative method is not very different from the experimental. Physiological experiments are designed to answer questions by letting you observe the immediate results to variables you provide. The comparative approach answers questions by letting you observe the results of variables that have been in place over millions of years.

The leaves are all down. They are brown now, but in the morning they will be tinged with white frost. As I sleep soundly in my shack in Maine, I will surely dream of tomorrow when I'll set out woodchuck, squirrel, skunk, and cat carcasses that I've collected (and frozen) over the summer.

NOVEMBER 1. It is still dark, and I'm already being awakened by raven calls! Several birds are flying over Kaflunk making short, high-pitched calls that are unlike the usual quorks. These calls convey excitement. The birds are flying to a kill! I *feel* it. Even I can understand, and I too am recruited. In an instant I'm up and out the door to note the callers' flight direction. Ten minutes later I'm off through the woods heading in that direction myself. A half mile into the woods I already hear the now familiar commotion ahead. *There*— the yells! Will it be a deer or a moose?

I come upon them down by Alder Stream. A dozen or so fly up and vanish in the thick woods, and as I peer through fir boughs down onto the stream, I see the skulking form of a coyote and the huge rack, hide, and skeleton of a monstrous white-tailed deer. (This rack was the largest I have ever seen, and it is now an officially registered Maine antler prize trophy that I keep as a pleasant reminder of the ravens.) All of the rocks protruding above the water in the brook are splashed by the ravens' whitewash.

There is not much meat left. The buck has been picked almost clean. The coyotes and birds must have been feeding here for a number of days. This is an animal that apparently got away from a poacher (hunting season has not yet started) to feed the ravens, instead, and to provide an excellent opportunity for an experiment. I will now test whether or not the shyness that I saw last March at the calf carcasses was indeed due to unfamiliar food. Surely all of these ravens are familiar with deer.

The skeleton and hide are still in one piece, and I drag it to the edge of the clearing near the log cabin. The ravens, missing their food at the expected place, should eventually find it at the new location. How will they behave after they relocate a food that is not only of a familiar kind, but also the same one already used by them?

Now comes the part that can be extremely boring and demanding: sitting and watching. I begin the task at noon. And, as always, it is a hard task, because I cannot do anything else, like reading. You can't see a raven fly by with your eyes averted.

Nothing comes all afternoon. It is a relief to go to camp at dusk, build the fire, and relax with a book. Tomorrow it should be different.

NOVEMBER 2. All day until 3:00 P.M. the remains are undiscovered. No raven, crow, or jay has flown nearby all day long. At three I decide to play my trump card. I position the speaker from my portable tape deck out the window of the cabin and play my recording of ravens' yells from last year. Within ten seconds a raven flies directly at me! I immediately turn the tape off, and the raven perches right beside me in the white birch tree, after circling the field once. This bird is obviously excited. It ruffles out its feathers, thrusts its beak in the direction of the carcass while making clicking sounds (beak snaps?) and barely audible murmuring noises. After fifteen minutes it flies off, not having uttered a single yell or a croak. Is this a selfish one that wants to keep the meat for itself?

Within five minutes I hear a quick high "queek" in the distance—an unmistakable yell—then again, closer. A half

dozen ravens suddenly circle all around, and more are on
the way! And more, and more, and *still* more! After fifteen
minutes I count twenty-nine ravens perched in the trees
near the deer carcass. I am spellbound, listening to a sudden,
incredible cacophony of yelling, croaking, and other weird
noises as the birds mill around. The surprise and excitement
nearly takes my breath away. These birds must have been
recruited by the one I just called in. Did it get the others
from near the site where the deer had been two days ago?
Maybe they had been gathered at a nearby roost.

Some might object and say that this is not really recruitment. But how would you demonstrate deliberate recruitment
short of asking the bird directly in Ravenese: "What do you
have in *mind* when your behavior causes others to share the
food? Is this what you *wanted?*" But I'm not asking these
questions. I'm trying to explain behavior *without* volition.
I'm concerned with the *effects* of the behavior, or how to
affect the behavior. What more do you need, when you see
someone near a bunch of kids deliberately jump up and
down, hollering an arbitrary set of sounds like "Candy,
candy," and then see the kids come running? Do you *really*
need to say, "Hey, kids, there is candy on the counter. Come
and get it." The meaning is seen from the *effect* on the
observer.

In other trials I provided only one carcass at a time. But
this time I had also planted road-killed cat, skunk, woodchuck, and gray squirrel carcasses in the vicinity, to widen
the net for potential discovery and to examine carrion choice.
Would all of these other baits now be consumed within minutes? No! All twenty-nine ravens act as if they are afraid
even of the deer that they probably recently fed from. They
seem to ignore the skunk and the cat, which are close by,
but they occasionally swoop and hover over the deer. Suddenly I hear lots of excited yelling, and a stream of birds
flies in the direction of the gray squirrel. A few minutes later,
other excited yelling from the direction of the woodchuck,
and a line of birds flies off in *that* direction. The yelling here
at the deer continues. What an uproar (and not all of it
sounds peaceful), but no raven lands at any of the carcasses!
Gradually, as dusk comes, the birds leave. None has fed here.

NOVEMBER 3. You can be sure I am on watch well before daylight. As expected, the ravens come at dawn—first five or six, and then at least thirty-five almost all at once. I sit for hours watching and recording their excited voices. But no raven goes down to feed all morning. They keep leaving and coming back, to feed finally in the afternoon. I wish I could have stayed another day, but I have to pick up all the meat and then leave for other commitments back in Vermont, after one of the most exciting weekends I've ever spent in the field.

So it was not the strangeness of the carrion that mattered to them as much as the newness of the *site*. And recruit they do; there is no longer any doubt. Furthermore, it is more clear now than ever that although the calls they make *at* a food source are a powerful attractant, recruitment also involves *another*, long-distance mechanism. The sound marks the precise site of the food, but for long-range recruitment the birds do something else that seems to include the scout actually alerting and leading recruits in. Whatever it is they do that results in a rapid buildup of birds at a food source, it is *deliberate*. But what is the intent? If it is to share the risk, what could possibly have been the risk at the familiar deer?

Ravens like company before feeding at an unfamiliar bait, and perhaps they recruit to have that company. But the fact that they would not land at a deer carcass that many of them had fed on just recently, even after forty were recruited, seems bizarre. They probably did not fear me or the cabin, because they had perched directly beside me facing *away* from me. They had shown no fear of the largest hawks—both the goshawk and the red-tailed hawks—that had come near them on several occasions. They were unconcerned about dogs, coyotes, and even a deer that walked near their feeding. Ah! I think I have it. They fear traps or lurking ground predators at the bait! If so, then perhaps they will go unhesitatingly to a bait high up in a tree.

NOVEMBER 15. I'm back to give the ravens a choice. A pile of slaughterhouse offal is spread out covering a weathered old board twenty feet up in the old pear tree at the edge of the clearing below the camp. An equal amount of offal is below the tree.

My alarm clock rings at 5:30 A.M., and I vault out of bed, light the kerosene lamp, smash through the ice in the wash pan to rinse my face, and heat up a cup of coffee. It is −13°F outside (and inside), cold for this early in the winter.

By 6:00 I am in the blind, that is, in the frame of my unfinished log cabin a quarter of a mile below Kaflunk. I'm facing the eastern horizon, which is now bright orange. The sky is dark blue above and green to the sides, grading to dark blue-black in the west. The last stars are still visible. A flock of small finches is already twittering in excitement as they feed on the white birch catkins in the gray dawn, scattering husks onto the snow below like pepper onto a white tablecloth.

By 6:30 the tips of my toes and fingers are already numb with cold. Clouds are becoming illuminated against the eastern sky, which is turning as pink as the plastic flamingos found in many front yards in Maine.

At 6:50 the sun pops up over the ridge, sending rays onto the snow and glinting off the hoar frost on the grass. Thousands of bright flecks of light twinkle everywhere. The shadows on the snow are blue.

Slowly, imperceptibly, the play of colors in the sky shifts again and repeats in the western sky. No bird found my baits. Not all day.

NOVEMBER 16. Today there is more excitement. At 7:45 A.M. a blue jay comes by and without even a moment's hesitation flies directly onto the bait on the board in the pear tree. It feeds for about five minutes, then calls loudly. Another jay answers; a duet of six calls and countercalls follows until the second jay comes to the tree bait. It feeds side by side with the first.

The two jays are obviously a pair. For the rest of the day they come and go together, spending much time side by side, hacking off chunks of fat and meat and caching them in the neighboring woods. In all they make 127 caches today, confining all of their foraging to the meat in the tree.

Throughout the whole day the two jays were secretive.

Aside from the duet following the discovery, one of the two called only (on six occasions) just before it left the area, and it was always away from the bait at the time. Apparently it called when it was leaving, possibly to bring the other one along. Indeed, after two or three calls from the woods, the other left the meat, and they flew away together. Given these details, it seems to me that their very infrequent calling was not meant to bring others to the bait, even though they have the capability to give and understand a "follow-me" message.

It is possible that unintended listeners could eavesdrop. But these birds showed that there are few rewards for potential eavesdroppers. After they had been caching meat for thirty-seven minutes, a third jay arrived, and was immediately attacked. The same (or another) bird appeared five more times. Each time it was chased by one or both birds simultaneously. Sometimes both chased the newcomer continuously for five minutes at a time, usually pursuing it for at least fifty yards. Apparently this pair had staked their claims, and they very effectively kept it to themselves, even though they took the precaution of caching liberal supplies of provender in the trees over an area of at least an acre. This aggressiveness underscores the unique sharing of the ravens.

Near mid-day, at 11:31, a pair of crows arrives. They give three to four caws, perch in a tree next to the baits for three minutes, and depart. They do not return today.

Part of the crow population migrates, and I have seen flocks of over a thousand in both the fall and the spring in Vermont. Usually they fly in small groups of a dozen or a few dozen, then they stop to feed before flying on. The crows who stay for the winter seem to stay in pairs, like these two. Presumably they are adults who are better able to feed themselves here than the younger birds that are forced to migrate.

NOVEMBER 17. At 7:18 A.M. a raven *finally* flies over the clearing. One might think it did not see the baits, but a raven could hardly miss such obvious targets exposed on the snow and in a tree. The bird makes a small loop before continuing on, silently. Throughout the day it comes back four more times, each time flying lower. Apparently it also appraised the situation while perched in the nearby woods. For about ten minutes I hear strange doglike growls, barkings, gasps, and grunts from the nearby woods. They say that if an experienced woodsman hears sounds that he doesn't recognize, chances are they are made by a raven. I did not know what the sounds that I was hearing were, until the interjection of a single hoarse croak gave them away. Why is the bird hiding? Is neither bait acceptable?

NOVEMBER 18. Ravens, it seems, are slaves to the familiar. As I had expected, the raven is here at dawn. But so is one other. Unlike their behavior yesterday, both now perch very conspicuously on the top twigs of a poplar tree north of the bait, and in only three minutes one of them swoops down and lands beside the bait. The bait on the ground! The bird stays on the ground for two minutes, then flies up and perches in a small sugar maple next to a white birch south of the bait, to join the first one who had just flown there.

I mention these specific perches because throughout the day usually one, but sometimes two ravens returned on eleven occasions, and when they came close enough to land (as opposed to just flying by), they used the *same* perches again. That different birds would use the same two perches out of the whole surrounding forest seemed hardly likely, and my suspicion that these were the same birds was heightened by the fact that they usually used the north perch first, then swooped over the bait or landed next to it (twice), and then used the south perch. After that (on four occasions) they departed in a beeline approximately 30° east of Mount Bald. That is the direction to Hills Pond, where a pair has been nesting every spring.

The ravens, though hesitant, do not seem greatly afraid of the bait. Perhaps they are simply not very hungry, rather than extremely cautious. One bird had already landed beside the bait at dawn, and at 1:30 P.M. a bird lands again. Before and after landing I hear the knocking sounds that I've heard before when birds are feeding or about to begin feeding. Why has the lone bird gone down to the bait without feeding, while the other stood watch? I had predicted, along the lines of the selfish-herd hypothesis, that, as it had been last March, *no* bird would approach the carcass until a critical

number had collected. I am wrong again. These birds went
down alone, and I now predict that they will not recruit.
There is no need to share now that they have no more fear
of the bait.

At 3:39, just before dusk, six ravens come by and circle
low over the clearing, apparently making a quick reconnaissance. Recruitment? Am I wrong *again?* I am very unsure
of what the ravens may do next, except that they will show
up again the following dawn. And *then* they will surely begin to feed.

NOVEMBER 19. Wrong! It's dawn. The pair arrives alone, flying directly from the east, and then perches side by side at its
usual place in the poplar. They are silent. But after seventeen minutes I hear a third raven in the distance. One of the
pair quorks loudly and flies off in the direction of the sounds.
The other one flies off as well.

At 7:40 A.M. one raven comes in silently and alone. After
ten minutes it flies down to the bait, pecks at it, and jumps
up and down in the style of dance I've come to expect at a
new bait. It keeps pecking and jumping, pecking and jumping, and then jumps less and pecks more. Strange. The ravens
seen here before had acted nonchalant. It does not look as if
this bird is interested in tearing anything off. But after eight
minutes it finally seizes a hunk and flies off to cache it. Then
it goes to work making more caches. It caches in trees and on
the ground, at distances of two hundred feet to perhaps over
a half mile from the bait. It goes about its work silently, and
I can see no other raven near.

After the fifteenth cache it suddenly flies off down the valley. Now I hear it yell. Is it recruiting? I can see it coming
back, a black speck in the distance, and about a quarter of a
mile behind it is another bird, coming directly toward the
bait. But the follower looks strangely light, not black like a
raven—it's a red-tailed hawk! The raven and the hawk both
land in the tree beside the bait and appear to pay little attention to each other. Did the raven see the hawk flying in the
distance and mistake it for a raven? The raven now resumes
caching, totally oblivious to the hawk's presence. The hawk
leaves after only five minutes.

After the nineteenth (and last) cache the raven revisits two previously made caches, as if checking its memory. Then it flies off into the distance. Again it calls. I can hear its calls getting fainter. Recruiting again? Within two to three minutes two ravens arrive. One bird hovers over the bait as if afraid, but soon it begins to feed, then it begins to call loudly, keeping it up for five minutes. It leaves, continues to call while in flight, and comes back two minutes later, still calling for another twelve minutes. Now it goes down to feed and to make three caches. Meanwhile, the other bird has been silently feeding and caching. Twelve minutes later I hear loud excited quorks: Three ravens circle and soon leave.

My four days of continuous watching are over. I had hoped to see the ravens go unhesitatingly to bait in a tree so that they might help me differentiate between two hypotheses. They were not so obliging. They clearly do not prefer to go to bait in trees. Given the same choice, the blue jays, in contrast, stayed in the tree, avoiding the ground baits even before any ravens were near.

The blue jays, it seems, have a call that says "Come here" or "Follow" to a mate. But they do not use it to recruit others to a bait. Instead, they persecute interlopers vigorously. Why should ravens behave so differently and tolerate and even recruit others?

To add to the confusion, I got a letter from Kathy Bricker, who had worked as a wildlife photographer in Michigan and who had been doing some raven watching on her own while waiting at a beaver carcass to film wolves. Her observations made me doubt that ravens really share at all. She wrote:

14 Ravens at 11 A.M. (7 seen yesterday) A lot of fighting ensued—no friendly play, that. They jab viciously at one another's head—I saw one grab hold of another's leg with his feet; the two thus joined stood with their beaks open, facing one another, and finally disengaged and flew off chasing each other. When they rush at each other with wings flapping, they throw out their feet with those sharp talons. I saw one knocked to his side by the blow; instantly he was surrounded by other Ravens who had been nearby. Apparently there is nothing quite so exciting as a fight, or perhaps quite so tasty as

raven meat. The meal in question righted itself and quickly flew off.

These observations are a bit disconcerting to me, because if ravens share food, they should not want to fight over it. I do not know what to conclude, if anything. Do they share information only on the location of food, and then compete for it with the others that come? It doesn't make sense.

WHAT IS ACCEPTABLE EVIDENCE?

A WOMAN FROM New Orleans who read my article on ravens in *Audubon* magazine, which I had written when I had just started to find out if and how ravens share, wrote me: "I did not have so much trouble as you did showing that ravens share. I see them at my feeder—they even feed one another." There are no ravens in New Orleans, nor anywhere else in Louisiana. Perhaps what she actually saw were several large dark birds (crows? grackles?), one of which fed another one or two (probably their grown offspring traveling along with them).

It is common to confuse personal interpretations with observations. It is a special bane in getting reliable observations on ravens, because there is so much engrained folklore about them that it is difficult to see them with a detached eye. But a trapper in the north woods might be more unbiased. I had read an article about a trapper/writer near Nenana, Alaska, and knowing he would be familiar with ravens in the north, I wrote to ask him if he had seen crowds of ravens at carcasses. I also told him the reason for my interest. He had a lot of raven stories to tell. First, he said "everyone" he knew, knew that ravens share their food. He was surprised at the ignorance of us armchair scientists so far away, who would even question it. He felt frustrated by biologists who didn't believe what people like himself who had lived in the woods all their lives knew. For instance, ravens "knew" the destructive power of his rifle, and they kept just out of its range. Nevertheless, they were "clever enough" to raid the fish he kept on racks for his dogs. They did this by posting a "twenty-four-hour guard" at his cabin. (How did he distinguish this, I wondered, from birds waiting for an opportunity to feed?) As soon as he got up, a raven was there to see him and "spread the word." (Read: Flew away, and/or called.) He claimed that one raven "followed" him all day. (Read: He occasionally saw a raven.) It then "reported back" to the others so that they could all leave just before he got back from his day on the trapline. (Read: He saw several leave together, and there were none when he got back to the cabin door.) Many of the birds "raided" (fed from?) his fish rack, and his idea of them "getting out the word" to ravens for miles around is that the one who discovers the food calls, and the summons is heard by all the birds in neighboring territories, who then also call, in an ever-enlarging ring of information sharing. (An interesting *thought*.) It was no mystery to him why the birds would do this: "gossiping." "It seems obvious," he said, "that the birds get excited, and they simply can't hold in their excitement that lets others know." (Very weak, incontinent birds, these.) And why should they evolve such transparent excitement? That, too, was "obvious": "Because it is best for the species." This stock answer explains nothing, any more than saying "God" created the world when you really have no idea.

It was disturbing to me to see someone so facilely blur the distinction between observations and interpretations and then even go so far as to make numerous deductions that he believed to be self-evident without the slightest shred of evidence for even one. When I was very young and didn't "see" what seemed obvious to adults, I often thought I was stupid and unsuited for science. Now I sometimes wonder if that is why I make progress. The ability to *invent* interconnections is no advantage where the discovery of truth is an objective.

There are those who believe that science consists entirely of disproving alternative hypotheses, as if when you eliminate the alternative views, the one you have left is right. The problem is there is no way to think of all the possible hypotheses that nature can devise. More than that, you have to prove which is the *most* reasonable. But any one hypothesis can, with a limited data set, be reasonable. There is at least a touch of truth in the idea that one variable affects another.

So if you look long and determinedly enough you'll find an effect on almost any variable you choose to examine. But I'm after the *big* effects, the central issues. To find those, I have to almost ignore the little ones. One has to be able to skim over—be superficial with—what is not important or relevant to your problem, and to concentrate long enough on the prime movers to unearth sufficient facts that, presuming they are recognized, add up to something.

The approach of seeing "blindly" to arrive at the truth was stressed by Sigmund Freud in a 1912 lecture to physicians interested in uncovering the unconscious in patients. Freud said:

The technique . . . consists simply in not directing one's notice to anything in particular and in maintaining the same "evenly-suspended attention" . . . in the face of all that one hears. In this way . . . we avoid the danger which is inseparable from the exercise of deliberate attention. For as soon as anyone deliberately concentrates his attention to a certain degree he begins to select from the material before him; one point will be fixed in his mind with particular clearness and some other will be correspondingly disregarded, and in making this selection he will be following his expectations and inclinations. This, however, is precisely what must not be done. In making the selection, if he follows his expectations, he is in danger of never finding anything but what he already knows; and if he follows his inclinations, he will certainly falsify what he may perceive.

Then there is the error of omission, which is inherent in almost all scientific papers. In this age of information glut there is a limit to how much written or verbal space-time you can commandeer from your colleagues who are obliged to try to keep up with what's going on. Having been fortunate to have time and resources to study the matter, a scientist must cut through the haze of details and get to the heart. You give the minimum of information—just enough to make the point. As a result, only a very sketchy (and often one-sided) part of any one argument is or can be presented. A problem arises when others' perception of the problem revolves around precisely those details that you may have pared. Those who touch the proverbial elephant on the tail may think you are uninformed—or worse, trying to avoid or cover up the truth—if you only talk about the head. Writing a book is therefore very satisfying to me because I can present more of the thoughts and richness and complexity of detail that make the "elephant" alive, interesting, and, I hope, also more real.

Animals provide inordinate intellectual challenges to understanding. But it seems to me that they are not just intellectual concepts. They are real things to touch, smell, hear, observe, and get close to. And if one does these things, one will see puzzles soon enough, which will generate even more observations, and which will eventually coalesce into concepts. Theory is like a scaffolding that you build up around the organism so that you can better observe its architecture and basic design. But after the organism has been revealed, the scaffolding is no longer terribly relevant.

There are different ways of seeing, and some people have a better "nose" for determining the nature of things than others. In the early part of this century, it was generally recognized that Konrad Lorenz knew more about animal behavior than anyone else. That made it easy. If *he* said that ravens have an innate call that they use when they want others to follow them, it was generally accepted as fact on the basis of his authority.

It is no longer like that. It is not seeing the truth that is now held important, but the proof or demonstration of it according to universally held, scientifically valid standards. So far in this study I have valid enough observations, but they are still too inconsistent to be acceptable evidence for any one idea or hypothesis.

EARLY WINTER CONFUSION

THANKSGIVING WEEKEND. As I drive past Hills Pond at dawn to come to the cabin (I was away last night), I see two ravens leave the pines along the pond where the nest was last year. So they sleep here. This is probably the same pair of birds that came from this direction to feed from the meat I left two weeks ago. The two birds fly together for about half a mile, then they separate.

I walk on up to the cabin to examine the fifty pounds of beef offal (twenty pounds of it is chopped suet) I put out last night at dusk. It is fully light now at 7:30 A.M., and I see or hear no birds. I don't really expect to see birds because now, near the end of the hunting season, they probably still have other food elsewhere, so I strike off into the woods. This walk is a luxury I will not be able to have for most of the coming winter.

The new snow from the last two days crunches lightly under my feet as I track a deer. But soon the track is lost among innumerable others. The oak-beech woods near Houghton Ledges are trampled like a barnyard, and heavily worn paths lead to the thick frozen swamps below. There are deer tracks everywhere, but not one deer is to be seen. I see no bear tracks. The bears are probably in hibernation. But there are many coyote tracks. The record of the snow also shows the paths of mice, squirrels, rabbits, weasels, fisher, and ruffed grouse.

By 9:20 I'm at least a mile to the north of the bait, following ever more trails. I'm alerted suddenly by a high-toned excited call of a raven, about a half mile north of me. I stop and listen and hear more raven calls, and then four fly over, heading directly south—toward my bait at the cabin! Sure enough, in a few minutes I hear the excited calling of birds flying toward good bait—the short, quick quorks and an occasional high-pitched queek. There is no doubt about it: Ravens are heading for a bait, possibly mine, and I rush to return.

It takes me forty minutes to get back to the cabin. The ravens have not only been there for some time, but they have already removed about half of the twenty or so pounds of suet. These birds had obviously not been afraid to descend to feed almost *immediately!* I am amazed and totally baffled.

By accident I stumbled on one piece of suet at least a half mile from the bait, covered by a small cone of snow heaped onto it. A cache. The raven tracks in the snow leading up to it give it away.

I go into the cabin and watch. The ravens seem not to notice me. Within a minute they are already back at the meat. As many as eight birds feed at any one time, but it is impossible to know how many there are here in total because they keep coming and going continuously. The big birds work side by side on the bait in seemingly perfect harmony. Once in a while one jumps up—it looks almost like an aggressive interaction or the result of one, but I don't think it is; I think it is just nervousness.

The sounds—the usual nasal quorks and much yelling and "knocking"—are music to my ears. One bird even flies around, making loops of a mile or so, and making high-pitched quorks.

As with almost every other experiment so far, the results of this one are a surprise. But they also fuel preexisting suspicions. Maybe some of the long delay in feeding that I saw before was because the baits were not tempting enough. Here they have *suet!* The quick recruiting to a very desirable bait supports the sharing hypothesis. Also, it is clear that at the rate these birds are lugging off the bait, there won't be any left in another day. Yet the vigor of their yelling is not diminished. The birds show absolutely no hesitation now in flying down to feed, and they are *still* noisy. This does not look like selfish-herd behavior.

As I go back into the woods, a mile or so from the bait, I continue to hear the birds at the bait, as well as those flying about in the immediate vicinity and those miles from it. If I were to assign an emotion to the sounds of the birds, I would call it a mixture of excitement and happiness. They seem to be irrepressibly boisterous. If I were a hungry raven and found such a "happy" bird, I would follow him no matter where he was, and I would soon be led to a good feast. I doubt if the birds do less. Given this noisy display, I doubt that there will be any of these swift and far-flying birds within an area of a hundred square miles who will not very soon be alerted to a good food bonanza.

DECEMBER 12, 1985. It already gets dark at 4:30 in the afternoon, and I don't make it to Maine to the hill until 7:00 P.M. The stars are bright against a dark moonless sky. Six inches of fresh new snow outline the silhouettes of trees, but I get no depth perception on the ground as I stumble up the steep hill carrying the heavy ANRS portable stereo cassette deck over my shoulder, the huge parabolic reflector and microphone in one hand, and a bag of groceries in the other. Suddenly on a very steep part of the hill my feet slide out from under me on a huge hard ice patch below a seep. My borrowed equipment goes flying, with the cassette recorder crashing onto the ice, spilling all six of its 6-volt batteries. Sorry, Dave! I'm afraid it is ruined. And I have no flashlight with me. But groping with my bare hands in the 20°F snow has some rewards: I recover all of the batteries. So far so good.

The bait has to be carried up next. This time I have brought along about four hundred pounds of suet that I have collected from slaughterhouses in Vermont. Suet is what they "instantly" recruited to the last time, so suet in superabundance I had to get for them. All of it has to be maneuvered up and put into place before dawn tomorrow. I carry a hundred pounds at a time in grain bags slung over my shoulder. This gets to be tiring up the steep slope through the woods along the steep ice field.

Someone had been walking in the snow ahead of me. Was it a hunter who came back, having found the cabin in the fall? Inside, on the table, I've put a note saying, "Make yourself at home, but please leave it as you found it. Thanks."

When I left the cabin last time, I used a twelve-ounce can of Meister Bräu to hold down the note. The can is still there, and another has been added: a can of Coors. Newly washed dishes are in the sink. The cassette tape recorder still works. The suet is all up before eleven. Content, I crawl into my sleeping bag covered with two blankets and a tanned deer hide.

As I awake at dawn, the temperature is 10°F inside the cabin and 9°F outside. The sunrise is colorful over the snow, but wisps of clouds soon creep in from the east in the stillness. An overcast is developing. Maybe it will snow again. The snow on the trees seems frozen in place. Nothing is stirring at the four hundred pounds of suet. I have placed a stuffed raven from the museum next to the bait in the hope of decoying the birds in even faster.

The silence continues. Nothing for hours. Then, at almost ten, a lone crow lands in a balsam tree overlooking the bonanza. It caws three times, sits there silently for another thirteen minutes, and departs. A half hour later a lone blue jay discovers the bait. It also calls three times, then remains silent and stays away.

DECEMBER 14. SATURDAY. The stillness has continued, except for a new sound—the barely audible hissing of falling snowflakes. It has snowed since late yesterday afternoon, through the night, and it is still snowing this morning. About a foot has already fallen. The white-mantled conifers are outlined against the gray sky and snow has coalesced into thick patches that press each branch onto the one below. And still the flakes keep drifting down gently, some clinging to the branches and the remainder filtering down to the forest floor. The constant downward motion of the veil of snowflakes mesmerizes me. And I wonder: How can the ravens possibly find food now?

By afternoon, outside temperatures rise, and the heat from my fire in the cabin helps to melt some of the snow on the roof. Soon a row of icicles sprouts at the edge of the roof and grows down in front of the window. They grow to about two feet and stop. A wind has come up, and temperatures drop from 25°F to 5°F within two hours. The weather change was signaled by a soft sigh emanating from the breeze through the trees, but soon I hear a deep moan that turns into a loud hiss, rising and falling in pitch. The storm roars, whining and raking off cushions of snow, thinning them and whipping them through the trees like white sheets. Bare pine branches flutter like black flames, fir trees nod, and birches sway and flail their thin dark branches. The wind propels itself down the chimney so that the bottom of the stove becomes the vent, and the cabin is soon filled with thick choking smoke. I am now forced to pull the remaining burning pieces of wood out of the stove and throw them into the snow to extinguish the fire. I always hate it when I have to do this at subzero temperatures. But I like it even less when it happens in the middle of the night.

There is nothing to do except to stay tightly wrapped up in my sleeping bag by the window and continue the watch. I cannot let my attention wander during an experiment. By dusk at 4:30, the cold has finally made me lethargic, and I go to bed. I'm fully dressed in thermal underwear, a thick sweater plus two shirts and wool stockings. My down sleeping bag is covered with two blankets. But I'm still kept awake by the cold as the temperature keeps dropping. By dawn the thermometer tells me it is −15°F in the cabin.

Listening to the howling wind and feeling its chill even inside the cabin, I marvel at how any bird can remain alive out there. It is amazing enough that the seed eaters can keep warm, but what about the insectivores—chickadees, nuthatches, brown creepers, and kinglets? It is a miracle. The ravens should be very hungry, both because snow has covered up other food, and because the cold will require them to eat more.

DECEMBER 15. SUNDAY. The morning stars twinkle brightly, the clouds have vanished from the sky, and the air is deathly still again. At dawn I hear a pine grosbeak, that is all. At 7:05 A.M. a pair of ravens comes by, as expected. But they do not go down to feed, and they leave after eight minutes. They come back five more times today. But they do not sound terribly excited, and they do not go down to the bait. Why? It seems very strange. Two weeks ago, when one would have expected no food shortage relative to now, a large number were already feeding a few hours after the suet was discovered. Is it possible that they are somehow aware of my presence in the cabin? If so, they should go down to a more distant food. I must test this immediately, so I put a second suet pile in the woods at double the distance from the cabin, with the stuffed raven placed on top.

DECEMBER 16. MONDAY MORNING, 9:01. An event: A single raven discovers the second food pile. *This* bird acts as though it is excited, flying around and making high-pitched caws for many minutes. At 9:16 I hear its calls receding into the distance. Unlike many of the others I have seen, this bird is an extrovert, and not a cautious one. I predict it is on its way to recruit!

It is. At 11:05 I hear raven calls from the woods. Several birds are in the vicinity. A good time to test the recruiting power of the yells again, so I play my tape for ten seconds. Magic! Within fifteen seconds *five* ravens swoop low over the cabin! But the birds don't stay near the bait; they return to the woods, as if to hide. I play the tape three more times in the next two hours when I see no ravens near, and it attracts ravens every time. Sometimes I hear their metallic knocking calls from the woods. Why don't they go down to *feed?* What in the devil is going on here *now?* The stuffed raven near the bait does *not* act as a decoy. Indeed, one bird finally comes down, not to feed at the bait, but to give the dummy a peck. This is strange. Very strange.

Could it be that they are *still* afraid of me in the cabin? During a lull when I hear or see no ravens around, I dash out and dump a pile of food even farther into the woods, far out of sight of the cabin. An hour later I check, and to my surprise the tracks in the snow show that a raven has not only landed there, it has also fed! And there had not been a dummy to help draw it in. Maybe I have kept them away after all.

On second thought, from all my previous watching of the birds, I cannot quite believe that they are afraid of me in the cabin. I enter or leave it so seldom in daylight, and then only if there are no ravens around. Maybe they prefer the thick

woods at the edge of the field? If such things make a difference, I cannot very well begin to get meaningful answers to
my experiments until I know.

NEAR CHRISTMAS. I stopped in on a sheep farmer in Hinesburg, Vermont, where I had seen two dead sheep lying out
in the pasture for about a week. Yes, I could remove them.
They had been killed by coyotes the week before, said the
farmer. Then he added, "Coyotes will strip a sheep and leave
nothing but hooves and bones in one night." Maybe. I have
learned to be very leery of the testimony of witnesses. Others'
testimony is not acceptable evidence. Not long ago I stopped
by at the landfill in Bethel, Maine, and witnessed the most
unusual sight I had ever seen. There were about fifty ravens
feeding in amongst the herring and black-backed gulls. I asked
one of the locals unloading a truck if he had ever seen this
many ravens here before. "Sure. They are here all year
long," and as I was leaving, he added: *"We* call them *sea
gulls."*
    A few days later, on December 24, I drove to Maine and
dragged the two sheep up. (This time there were no black
"sea gulls" at the Bethel dump when I went by.) The deep
snow made the trip up the hill quite an effort. I skinned one
sheep and cut it open, leaving it well in the woods rather
than at the edge of the field, at least one hundred and fifty
yards below the log cabin. The other one I brought up to
Kaflunk to keep in reserve. I covered it with an old rug rem-
nant so the coyotes would not eat it, and drove down to join
the family for Christmas Eve.
    Late after the Christmas Eve festivities, I was glad to be
able to come back up. It was a dark night, and cold. My family had said, "We're glad *we* don't have to climb up there
tonight." But to me the physical challenge was more than
compensated for by the thought that I might learn something
about ravens in the coming dawn.
    This time I did not want to watch from Kaflunk, although
the possibility of a warm fire there made it extremely tempting, because I was still not sure if the birds refused to feed
the last time because they were afraid of the cabin. Or could
they smell me inside? To test for the first possibility I had

left meat within fifteen feet of the cabin. This was all gone now, and as I scraped away the new snow, I found an abundance of raven feces splattered on the packed old snow—acceptable evidence that ravens have fed here. So they appear *not* to have been afraid of the cabin after all, but I still don't believe that it's true.

DECEMBER 25. The morning is very foggy, and I begin my vigil as it is getting light near seven. Through the dense fog I hear two ravens quorking, then I hear excited, high-pitched, odd-sounding calls. Then it is totally quiet. The fog lifts, and it starts pouring. A raven comes over during the deluge, banks over the sheep, and continues on to make a two-mile circle, while calling in high-pitched quorks. Will it bring others?

A half hour later the rain turns into a snowstorm. A heavy one. In a half hour the sheep would be buried. No use to watch it now, so I cover it up again with the rug and return to Kaflunk. And there, *directly* in front of my doorstep, in my own footsteps, are fresh raven tracks! The human scent about the cabin had *not* deterred this bird. Nor did it fear my footsteps. Two of my worries are much reduced. Maybe it was the smoke from the fire that had kept them away the last time? I will have to go without a fire until I find out.

DECEMBER 26. The night was quite cold. Temperatures in the cabin dipped to −10°F. I had pulled myself deeply into my sleeping bag, as if to hibernate. The cold made the clock slow down—it lost half an hour, and so I didn't wake up until 7:10 A.M., when it was already light. This panics me because one of my ongoing experiments is to see whether or not ravens coming in crowds are preceded by individual discoverers. Since a *discovering* raven may merely fly silently by, I need to watch *constantly*. One lapse of ten seconds can spoil an experiment that lasts weeks. This is a lapse, and I'm angry at myself. But in a few minutes I am dressed and running down through the snow to uncover the sheep.

Three times throughout the day a raven flew over. Despite the sheep being deep in the woods, the raven made no attempt to feed. Instead, it pecked for half an hour on a dry old rib cage of a deer sticking out of the snow in the middle of the clearing directly in front of me by the cabin. This proves one thing: The birds were *not* hesitant to come down to feed at the sheep because I was in the distant cabin (usually watching through cracks in the yet-unchinked walls, or through a one-way mirror in the window). Do they fear the bait itself? But why?

Near midmorning the cold finally overwhelms me. I rush up to Kaflunk to build a fire and have a cup of hot coffee. A surprise awaits me: Four ravens fly up, and there are tracks all around the cabin. Indeed, the birds have walked within three feet of it. They had been silent. I will now put plenty of wood on the fire, to get a good plume of smoke coming out the chimney. Will that keep the birds from coming back?

In the evening as I come back, after having left the snug fire, I find no new raven tracks around the camp. So, by process of elimination, it seems that it must be the *smoke* that keeps them away! I feel good about my detective work, but not about the implications.

DECEMBER 27. At 7:15 A.M. a raven flies over. It is excited, making high-pitched quorks. The sheep carcass in front of Kaflunk has been discovered. This raven flies over repeatedly, making huge loops over the valley and toward Mount Bald and Gammon Ridge, giving continuous, high-pitched quorks. No yells. Strange. It does not feed. No more birds come. I do not seem to be getting anywhere this session, either, except maybe to find out that the birds are indeed afraid of *strange* bait.

DECEMBER 28. The fourth day. No ravens this dawn. I build a fire and wonder whether I should stick with it another day, when at 9:16 A.M. I hear high-pitched quorks in the distance. In a moment two ravens are circling nearby. Then the *noise* starts. First there is a total of forty-six high-pitched trills. This is very unusual. At 9:24 one raven begins to feed at the sheep, to be immediately joined by a second. Both are silent until 9:47 when feeding stops. From the woods I soon hear three more trills and then seventeen high-pitched yells. Why don't these birds join the feast with the two? Later one of the

birds is perched in the big birch next to the cabin, serenading
with wild abandon, a mixture of grunts, groans, gurgles,
moans, screeches, yells, and croaks with all the gradations
between. After a half hour I do not hear this repertoire
again, only occasional yells. And this morning I do not hear
the knocking sounds. I have a feeling something is going to
happen now, and I must stay to find out. Again, these observations are not only inscrutable, they seem to contradict
every hypothesis I have!

Watching. Dreaming. Wondering. There, through the
spruces, on the ledges about fifty yards away—standing stock-
still and looking at me—a coyote! It looks like a dark yellow
German shepherd. It looks for several minutes, then melts
back into the trees. A raven circles over it and croaks ener-
getically. I have no way of knowing whether the coyote was
attracted by the calling of the ravens or the scent of the meat.
(I later checked the snow around the other, unwatched sheep
where no ravens were feeding, and the coyote had not been
there.)

It might seem too costly for the ravens to be noisy near
their food if this attracts mammalian scavengers. However,
in the last six weeks we have had six snowstorms, any one of
which could have buried the bait. A coyote feeding at the
carcass would not be deterred by the deepest snow. It would
keep uncovering the bait. The ravens might not be able to do
so. It seems logical to me that the ravens might need the coyote, to kill the prey, to open it, and to keep it uncovered. I
presume that in these winter woods, the ravens need the
carnivore.

DECEMBER 29. At dawn I can already hear high-pitched
quorks. Four ravens arrive simultaneously, and now I hear
three series of knocks from the forest. Then it is quiet. After
twenty-three minutes two ravens go down together, to be
quickly followed by two more. There are lots of yells, but no
trills this time. After a week in the woods, I'm no wiser.
Sometimes they recruit. Sometimes they don't. How many
variables are there?

ARE RAVENS HAWKS OR DOVES?

I T MAY SEEM a matter of idle interest to be concerned about whether single birds or groups discover baits, to try to find out which individual birds are at a bait at any one time, how long they stay, whether or not they return, and whether or not they call or fight. But the answers are relevant to the more important questions of whether or not sharing occurs in addition to recruitment, and which mechanisms possibly allow for the evolution of sharing. What is at issue here is whether raven crowds at baits are social bands. If they are, then the theories that the birds merely help their relatives (kin selection) or those who help them (reciprocal altruism) must be examined closely. If the birds do not form social relationships, then both of these theories are moot, and others deserve our detailed attention, instead.

It is possible, however, that a raven who has neither friends nor relatives still finds advantages in cooperation. One way of looking at the problem is to say that the bird is in a "Prisoner's Dilemma," a decision-making theory espoused by game theorists and political scientists. This was named after a hypothetical situation in which two accused criminals who will never meet again share a secret about a crime they committed. Each is given a one-time chance to confess and squeal on the other in exchange for a lighter sentence for himself and a heavy sentence for the accused. If neither talks, however, they could both go free. Should one trust the other and try for freedom? Or should he "defect" and try for a lighter sentence? Applying the analogy to the raven's case, "talking" benefits the other and is the reverse of squealing. If two ravens meet, one could decide to tell the other where there is a carcass, in the hope that that other in return will divulge information about a carcass it knows or will know about. Sharing could get each bird more food in the long run whereas not talking confers the immediate advantage of keeping the carcass for itself.

It is not, of course, necessary to assume any thought, planning, or consciousness in the raven's or any animal's behavior patterns or "strategy." The behavior could be conscious or planned, but it could also mimic a rational response because animals with such responses have reproduced more than those with alternative reactions. The end result is based on mathematical probabilities, but we see *only* the result and have to decipher the equation that brought it about. Where one behavioral program competes against another, some theorists contend that over time the only solution to the Prisoner's Dilemma is mutual defection.

The situation changes if we consider the population versus the individual as the unit we wish to serve or benefit. In a world of very few carcasses or other super food bonanzas that are easily divisible among a number of individuals, it would generally be advantageous for each raven if all ravens were programmed to be "nice" and to share. Indeed, it would not be very costly to be nice if the carcasses were going to be eaten anyway by attending carnivores. So if, from the population perspective, it makes eminent sense for ravens to share, what reason is there to suppose that evolution might *not* have made individuals do so?

John Maynard Smith, a behavioral ecologist at the University of Sussex, formalized the argument in 1974 in the famous "dove-hawk" analogy in his theory of evolutionary stable strategies. In this analogy an animal can be either a "hawk" (defined as one who always escalates until it either wins or is seriously injured) or a "dove" (one who never escalates and, if its opponent escalates, then backs down). The contestants fighting over food (or over any other resource) never know beforehand which strategy their opponent will use; hawks and doves look alike. Clearly the individual "should" be a hawk when the chances are great of confronting a dove, because there is little danger of retaliation. But when the chances of facing a hawk are great, it is best to be a dove to avoid likely injury. Since the relative value of either strategy depends on the frequency of encounters with the other, neither hawk nor dove can exist for long by itself. For example, when doves are plentiful, hawks will enjoy a huge advantage and thrive, because they seldom get hurt when attacking and their opponents usually back down. Conversely, when hawks are common, they meet each other frequently and are killed or injured frequently, and doves (who always back down from a hawk but not necessarily from each other) have the advantage of not being killed or injured in an encounter. Therefore, in the long run, the evolutionary stable strategy—the "rational" (mathematical) strategy or the one where there is no mutant strategy that gives a higher Darwinian fitness to the one adopting it—is for *both* behaviors to persist in the population at the frequency where costs (injury or death) and benefits (rewards from winning a contest) reach a compromise that ultimately maximizes reproduction.

As in the Prisoner's Dilemma, in Maynard Smith's hawk-dove analogy any two individuals face each other only once, so there are no long-term consequences to weigh, because no contestant knows beforehand whether the individual it encounters is a hawk or a dove strategist. (If a protagonist *knows* it is or will be facing a dove, it should always be a hawk, in the same way that a raven who faces a non-reciprocator should never share.) The question then of whether ravens are hawks or doves (in this case, non-sharers or co-operators) might be resolved entirely differently if the animals interact repeatedly, recognize each other, and remember how the other behaved before. That ravens can recognize each other as individuals and that they have long memories are not points of contention. Many birds, especially corvids, have been shown to be able to recognize each other and to have good memories. Thus if, as indicated, individual ravens also associate in bands or groups, at least the prerequisites are in place for reciprocation, because one may therefore receive a favor in return for granting one.

Robert M. Axelrod, a political scientist at the University of Michigan, in 1984 wrote a book about alternate non-kin-based sharing mechanisms, *The Evolution of Cooperation*. He considered the results arising from different hypothetical strategies that contestants might use and then tested the strategies, not through evolution as is normally done in Nature, but via computer tournaments where one strategy could be pitted against the other within the computer, which would then determine the survivor, or the "best" strategy resulting from the selection of one versus the others. He ran fifteen strategies submitted by different game theorists, and after the survivor had emerged, he advertised a second tournament in professional journals and magazines for personal computers, telling the contestants of the results from the first round. Sixty-three more programs to challenge the survivor were submitted, and a second round-robin was held on the computer.

The winning behavioral strategy, submitted by Anatol Rappoport of the University of Toronto, was this: Automatically cooperate (share), giving the other the benefit of the doubt, and then do whatever the other did on the previous move. In this, the "tit-for-tat" strategy, cooperation breeds on itself. It is neither a hawk nor a dove strategy. It is both. Although a dovish individual might get bested in the first encounter, it is never exploited by the *same* hawk twice.

The message of the tit-for-tat strategy is that being "nice" is far more successful than being "mean." According to the computer simulation of the "meanest" versus the "nicest" possible strategies, nice guys finish first if they retaliate promptly but do not have too long a memory, and if they never try to get ahead by trying to get the other guy down. Is this how ravens behave? Perhaps. But the whole show is off if (1) they do not *repeatedly* interact as individuals (2) who identify one another and (3) have very long memories that extend at least to the last time the individual showed a potential reciprocator a carcass. It seems a lot to ask for, even in a raven.

One problem (at least in my mind) with reciprocation or tit-for-tat as a mechanism for carcass-sharing in ravens is that the *opportunities* for reciprocating are likely to be very few because carcasses are rare. Even if ravens form groups, it might be more than a year before a bird who is willing to reciprocate will be able to. This scarcity of opportunities to reciprocate increases the uncertainty of distinguishing genetic reciprocators from non-reciprocators on a one-to-one basis. One way around this would be if related birds (who may or may not know each other) stayed together. In terms of the mathematics of gene counting, helping relatives is almost the same as helping oneself. But ravens breed solitarily, and the young disperse (see later chapters). From all I had seen so far, long-term associations of raven relatives seemed extremely unlikely.

Cooperation is a very slippery concept to tie down. It depends on the bounds of the population of potentially cooperating or "nice" individuals that one incorporates into the model, on the costs of being nice, and on the delay or uncertainty that can be tolerated to get paybacks. Given these fluid uncertainties, I am somewhat skeptical of mathematical formulations, precise or otherwise, in determining the balance of any one situation. Perhaps ravens could have a "hopeful reciprocity" without any of the complexities of direct reciprocation as prerequisites. This model is based on observations of human behavior: I have noticed that in relatively sparsely settled regions motorists who are total strangers often blink their headlights at me when a highway patrolman is lurking up ahead scanning his radar in search of speeders along a safe stretch of highway. But they never give such warning near the cities. I have found myself reciprocating to other motorists. Maybe I do it because the cost of being "nice" is so low and the benefits of having like-minded people out there so high (i.e., keeping, as opposed to losing, one's license).

Maybe the same thing applies to ravens. If a raven gives up almost nothing by sharing a superbonanza and thereby gains even a small chance of the favor being reciprocated by some other bird when it is starving, then the skewed ratio of costs versus benefits may promote or facilitate the evolution of sharing behavior. Finally, the breadth of community or "group" from which a reciprocator might come is less a matter of closeness of kinship than of the breadth of magnitude of the external threat. The USA and the USSR would be instant friends and cooperators if there were a perceived threat from aliens from outer space. (How about the threat

of overpopulation and environmental degradation? Is that
threat not big enough?) But such threats are outside the pale
of experience and natural selection; they can unite only if
they are vividly perceived. Are ravens intelligent enough to
perceive that their common threat of starvation can be met
by cooperation? The possibility cannot be discounted. But
given no other evidence, and only one lifetime and limited
resources, I would explore the most plausible (or easiest)
hypothesis first, and the least, last. So far the kin selection
and reciprocal altruism hypotheses are not ruled out, but
they seem long shots. Yet none of the other obvious ideas I
can come up with appear to be strong contenders, either. Is
there then something totally unexpected operating in ravens?

RAVEN INTELLIGENCE

ROM THE ANCIENT NORSE, who considered the raven to be a messenger of the gods, to the Native Americans, who respected him as an all-knowing trickster and a god, to Konrad Lorenz, the common raven has been credited with being the most intelligent bird in the world. The perception of the raven's intelligence (i.e., conscious foresight or insight) as compared to that of other birds is summarized by W. A. Montevecchi, a behavioral ecologist who observed ravens at a colony of herring gulls he was studying. He told me that the gulls, in comparison to the ravens, acted like "vegetables." Nevertheless, despite the overwhelming force of consensus, there is surprisingly little objective evidence that can be used to compare the ravens' intelligence versus that of other corvid birds. Indeed, a review of the literature convinces me that no proof of the raven's singular intelligence has yet been published. Instead, many of the anecdotes purporting to show this are significant *only* if the intelligence is taken for granted from the outset.

The ancients presumed the raven was all-knowing. The historian Thucydides even attributed to the raven the sagacity of not feeding on animals that had died of the plague. (Thucydides did not say if animals that died otherwise were *also* not immediately eaten.) Similarly, the great Roman naturalist Pliny described as an example of great ingenuity how a raven contrived to quench his thirst: A raven saw water near the bottom of a narrow-necked vase, and to obtain it he threw pebbles into the vase until the water was elevated to within reach. Now, if the bird had indeed visualized what

111

it was doing, this act could be construed as showing intelligence. But it also seems possible that the bird was simply "caching" nearby objects into the jar. It then drank when water miraculously appeared. Would it also drop in objects that float? Would it also drop in stones when *not* thirsty? Thucydides lived in the fifth century B.C. About 2,400 years have elapsed, and to my knowledge, the simple tests of his assumption have yet to be made.

The raven's intelligence is commonly assumed at the outset when ravens work in "teams." P. J. Johnson, a resident of Whitehorse, Yukon Territory, who campaigned successfully to make the raven the territory bird, wrote to me: "It's a riot to see them [ravens] stealing from the huskies. Often one or more ravens will work together. One will attract the attention of the husky, placing himself just past where the limit of the dog's chain will allow him to tease him. This works as a wonderful distraction as the partner raven merrily perches on the dog's dish and eats his fill." Similar stories of ravens versus huskies or many other animals are legion. But that neither proves nor disproves that ravens consciously and strategically cooperate, planning and executing specific strategies. More than one raven may gather simply because they happen to see the same food. Perhaps one bird does indeed distract the predator, but without any conscious intent of creating a diversion. One or more of the others then merely takes advantage of the opportunity as it arises. Neither example is acceptable evidence for *showing* intelligence.

One published example of presumed "deliberate behavior" concerns a raven hunting mice. The bird was perched above a lawn covered with snow, then "glided down slowly and on landing immediately jumped around several times keeping its wings over its head." After this brief jumping spree, it returned immediately to its perch and studied the region it had just come from. It did this several times before flying away without apparent success. However, the jumping had broken open many vole tunnels and "it could only be concluded that this particular behavior was a deliberate attempt to flush voles from their subniveau hiding place." There is, of course, a much more plausible alternative explanation. As I have indicated, ravens routinely jump in the manner described above near *any* unfamiliar objects or prey, and without reference to snow or tunnels in snow, or prey to flush.

The ornithologist Thomas Nuttall observed that ravens carry nuts and shellfish up into the air and drop them onto rocks, and that these are "facts observed by men of credit." But is this behavior due to the ravens' singular sagacity?

This phenomenon is also well documented for birds such as crows and gulls, which are of lesser intelligence. Bent describes ravens playing with fir cones and other objects that are dropped and then chased in play. Perhaps a bird initially picks up an item known to contain food and then drops it in frustration or simply because it is bored with carrying it. With luck, the object might hit a ledge, and the bird then observes that food miraculously appears. Birds are known for their excellent memories, and such results could positively reinforce memory with no intelligence involved. Of course, it is entirely possible that one species may short-circuit the entire trial-and-error process by direct insight or intelligence. The significance of knowing by *insight* that you have to drop the shell from a certain height onto a hard rock in order to break the shell is obvious. But whether or not certain behavior is based on intelligence cannot be assumed.

Two single observations of ravens dropping objects onto human nest intruders and onto incubating birds have been considered to be tool use. But corvid birds habitually hammer objects near them when they are frustrated or angry, and loosened bits and pieces may simply obey the law of gravity.

In the one example purporting to show the raven's intelligent use of rocks, a raven at the top of a cliff dislodged rocks with its bill. The rocks fell down toward the intruders climbing to the nest below. This behavior is entirely consistent with what I see some male ravens do routinely in Vermont and Maine when I near their nests in trees. (At some nests both parents are silent and leave when a human comes near. At others both parents stay to fly around and call. At still others only the male stays to express his anger and frustration at the human intruders.) At one nest in Maine, where the female routinely leaves when one comes near, the male seems to go beserk with anger. On the dozen or so occasions that I've been near this nest over the last three years the re-

sponse is always the same. The male makes angry rasping
caws and other alarm calls, and he pecks with tremendous
vigor at whatever branch he perches on; and then continuing
his angry growling calls, he snaps at all twigs near him and
rips them off. There is thus a rain of twigs and foliage falling
to the ground almost wherever he lands, but it is unrelated
to the position of the intruder. Indeed, due to the many trees
there is no shortage of places to perch, and nobody has yet
been hit by one of the snipped twigs drifting down from above.

When one climbs the nest tree and feels the vigor of
his response, there is no mistaking this bird's power and de-
termination. But one does not feel that he is using his head.
Instead, one gets the impression that he is acting like a maniac. Every note of protest from the young (when we band
them) results in instantly renewed attacks on the branches
within reach of his bill. There is no hint of deliberate tool
use, which is probably as far from this bird's mind as it is
from a maniac's who smashes the furniture because he is
angry or hurt. Undoubtedly, if this raven were nesting on a
cliff and could perch only on the top, rather than on trees, he
would also vent his anger at the substrate at his feet. No
strategy need be assumed.

Many interpretations about the ravens' behavior are colored by the assumption of its intelligence. Zoologist Donald
Griffin makes a compelling and scarcely challenged case for
awareness being beneficial, and the detailed observations of
behavioral ecologist Eberhard Gwinner in Germany, showing
the varied ways in which ravens protect their young from
both high and low temperature extremes, look very much as
if the birds "know" what they are doing. But it seems from
the published literature so far that the raven's intelligence
has been little examined. Its cousins, the jays and crows, have
so far received at least as high marks for intelligence. For
example, a Northwestern crow, *C. caurinus*, used a small
stick to pry a peanut out of a crack. Similarly, common blue
jays, *Cyanocitta cristata*, were adept at using tools to retrieve
food. A New Caledonia crow, *C. manaduloides*, used a twig
to probe into a hollow stem. Captive rooks, *C. frugilegus*,
were able to select the proper holes to plug to retain water
to drink. But one can perhaps stretch the use of "tools" and

"intelligence" a bit. There are also reports of crows using automobiles as "nutcrackers," and jays using the branches upon which they place acorns to hammer, as "anvils." But one critical study on the American crow *C. brachyrhynchos* does not distinguish this bird as a genius in laboratory experiments. Rather, it showed behavior "comparable to that of pigeons, rats, and monkeys." However, given the tests, I might not have done much better myself. My personal impression is that the raven is very intelligent indeed, because many of its actions suggest awareness. But impressions are not evidence.

The following two anecdotes seem difficult to explain in terms of mere learning or blind stimulus-response mechanisms. The first concerns a pair of ravens I flushed while they were feeding on a large chunk of suet. The usual procedure of ravens, crows, blue jays, woodpeckers, chickadees, and nuthatches is that birds take bite-sized portions, tearing off protruding strips or corners. In contrast, one raven that I flushed from a large frozen chunk of suet had chiseled a groove 3 inches long and more than 0.6 inches deep around one corner. Given more time, it would almost certainly have succeeded in removing a much larger chunk of food than it could have secured by direct pecking. Pecking off what is immediately available is fully adequate for all other birds so far examined, so why not for a raven? It could be argued that the bird found it *easier* to take bites out of a groove. But this seems a doubtful explanation, given that small bits of fat adhered loosely to the edge of the groove. Did the bird show conscious foresight in order to get the larger piece?

The second anecdote involves caching behavior in two raven pairs in the wild. In the first instance, a bird feeding alone (later joined by its mate) hacked off bits of meat from a frozen carcass and *stacked them in a pile* next to where it was working, something I have never seen in any other species of bird. After the raven had accumulated a pile of approximately 15 cubic inches, it filled its gular pouch with more meat from the carcass, *then* picked up all the pieces in the pile of meat in its bill and flew off with its prize. The same bird or its mate repeated the process eight times within the next hour. (The two were at the bait together and neither took the other's meat.) Once while one of the pair was making a meat pile, a third bird came, perching nearby, obviously observing the pair remove meat. Finally it flew down, walked around the bait until it was directly behind one of the pair then working alone, cautiously advanced, and grabbed the loose meat pile. After that when several ravens were feeding at the same meat pile, I never once observed any of them piling up meat they had torn off.

Tony Angell quotes a similar observation in his 1979 book, *Ravens, Crows, Magpies, and Jays*. A tame raven was being fed crackers, and after eating its fill, rather than flying off separately with each one, it took a half dozen one by one and placed them side by side in a stack in a snowbank. It could then grasp the whole bunch at once, and it flew off with it. Raptor biologist Rick Knight at Colorado State University told me that he also observed this behavior by a pair of adult ravens in the Grand Canyon during January 1988. He and a friend provided the ravens with Ritz crackers, which they immediately stacked and carried off in piles of three and four in their bills.

I am not suggesting that animals, even ravens, cannot be programmed to drop loose food now in order to have more later, or to stack food before carrying it off. However, it seems unlikely that ravens *specifically* would be genetically programmed to make little meat piles to faciliate caching or to chisel out large chunks of food to make it possible to take more per trip. If it is blind genetic programming, then chickadees, nuthatches, crows, woodpeckers, and blue jays should just as easily evolve to do the same. Indeed, since they cannot defend meat piles against ravens, they should have *more* reason to have evolved these behaviors as a means of making the best of the limited time they can expect to have at any food bonanza. It seems therefore difficult to avoid the conclusion that ravens have, for birds, unusual awareness of both the consequences of their own actions and the anticipated actions of partners and competitors.

SHORT WORK OF TWO SHEEP

As I cam' by you ould house end
I saw two corbies* sitting theron,
The tane unto the t'other did say,
"O whare sall we gae dine the day?"—

"Whare but by yon new fa'en birk,
There, there lies a new slain knight;
Nae mortal kens that he lies there
But his hawks and hounds, and his ladye fair.

—Anonymous,
From an old Scottish ballad

JANUARY 2, 1986. I have been groping for an explanation of why there has been so little recruitment the last two times. Maybe there already is sufficient evidence to explain it, but I'm unable to see it. Maybe they are so smart and their responses so complex that I will never see it, unless I understand the individual birds themselves. Naturally, I hope and search for general rules, but these may remain hidden unless I can identify individuals or classes of individuals.

The most recruitment I have seen has been in the fall. And the greatest number of ravens I have seen together was at a landfill in the winter. Aha! I think I have it. In the fall the birds are still dispersed, but throughout the winter they

---

* Corbie *is the Scots word for raven, borrowed from French* corbeau, *which comes from Latin* corvus, *or crow.*

collect at permanent food sites, such as dumps. So there are fewer and fewer out here in the woods. Therefore, even though the few birds left try to recruit, they have no luck. One of the reasons I am going to Maine today is to reassure myself that no more than the four ravens I saw last week have turned up at the sheep carcass. You can never be sure.

As I go by the Bethel dump, it is 2:15 P.M. Strangely, not a single raven is in sight. Last week there were probably close to a hundred.

Going up the hill, I walk first of all by the log cabin. The sheep I left out in the field has not been touched. No raven tracks, no coyote tracks. But as I walk farther up the trail, I hear ravens at the sheep by Kaflunk! I retreat quickly into the woods to wait until dark so that I can enter the cabin without being seen.

What a surprise awaits me when I get there! There is nothing left of the sheep except bare bones, hide, and the frozen-solid stomach contents. The half acre or so of snow surrounding these remains is trampled by raven tracks. Were there *hundreds* here? Many dozens at the very least. Scratch another idea. There were plenty of ravens here, and it is mid-winter.

JANUARY 3. I awake to the beginning of a heavy snowstorm. One raven comes and makes a long series of quorks, then descends to pick at the bare ribs of the sheep. In a few minutes I hear the short abbreviated quorks I have so far heard only when several ravens are flying together toward a bait. Is it a "follow-me" signal? The sounds come closer, and four ravens swoop over and around the clearing. All is quiet. In a few minutes four ravens descend and pick at the bones.

After they leave, I cut a two-inch hole in the plastic sheeting in one corner of a window, behind which I mount my camera on a tripod. Later, when a raven comes, it does not croak. Instead, it makes the knocking call. No birds descend. Was the call a warning? Birds come by repeatedly throughout the day, but none of them lands near the bait. The snowstorm continues unabated.

Within a few hours there is nothing to be seen of the remains of the old sheep, or of the new one I put out last night twenty yards from the old. Snow covering up food must be a serious threat to these birds. So far every bait I have put out this winter has been covered several times, and one of my constant jobs has been to brush off the snow so that the meat can be discovered by the birds. A coyote might be handy to have around, to dig the carcass out whenever it gets covered by snow, which seems to be every several days. But if *that* is why the recruiting behavior has evolved, the ravens should be calling *more* now while it is snowing, rather than less.

In Vermont, I have been watching a single pair working on a bait I have set out within sight of my bedroom window. The birds come every morning within a few minutes of 7:30 and stay for about an hour and a half, coming back for another visit in the early afternoon. The bait I have provided is a huge pile of suet, most of it in two- to six-inch strips. The two birds *could* eat a piece in a few minutes and then leave. Instead, they pick, pick, pick, spending minutes on the pile before finally flying off with a piece, presumably to cache, and then coming back to repeat the process. It seems to me that this constant picking keeps the meat exposed. Otherwise, if it snows while they are not there, it might vanish from sight. The more birds there are, the more likely it is that some suet will be there to pick. Is this why recruitment has evolved? If it is, then they should not have recruited at the moose in the fall, or when it was *not* snowing. This idea doesn't hold either.

JANUARY 4. It stopped snowing in the night, and the wind came. As usual, I wake up enveloped in choking blue smoke as it pours down the stovepipe. Time to throw the smoldering logs out in the snow again.

It is just a degree above zero in the cabin. The wind is howling fiercely under a dark blue sky studded with stars as I shovel the snow off the sheep and rush back inside to melt some over the Coleman stove for coffee. (I won't build a fire in the daytime because the smoke might alarm the ravens.) I am very hopeful this morning. All of the ravens' food supply, even if they have caches or other sources, is likely to be deeply buried. They will have to come here.

Indeed, at 7:33 A.M. two land in the birch near the bait,

and three more fly over. Only one stops. For twelve minutes
the lone raven makes groaning noises, snaps its beak, croaks
softly, and trills loudly. I have heard this song before. It
means there are others near, and they will probably soon
begin to feed. Are they hiding in the woods? Suddenly, in
five minutes, there are six to eight. At 8:16 there are three
at the bait, then five. For the next hour there is constant
traffic of birds to and from the sheep, with three to five always feeding at the same time. At about 9:30 there are noticeably more. By 11:00 I usually see ten to fifteen at the
same time. At 11:30 the numbers have increased to at least
twenty-three! In the afternoon they taper off gradually, and
at about 3:30 P.M., near dusk, only ten are left. By 4:00 they
are all gone. Now I can leave the cabin and examine the
damage. The new sheep is at least three quarters gone! And
throughout the woods I see tracks on the snow where they
have landed and cached meat, covering it with an inch or so
of snow.

Most of these caches were later retrieved by coyotes. But
the ravens may need the coyotes, too, to find snowed-in carcasses. Earlier I had tested whether or not the ravens were
able to find or retrieve meat buried in snow at other than the
expected location. Two fresh pork chop bones, one freshly
killed red squirrel, a heap of frozen sheep entrails, and one
run-over cat were buried under two to four inches of freshly
fallen snow at different locations within one or two yards of
where twenty ravens had been feeding at the sheep. I then
removed the sheep. It was snowing lightly when the birds
returned, and several of them started pecking into the snow
and making sideways shoveling motions with the bill at the
precise site where the sheep had been. Eventually one bird
contacted a foot of the cat, and the whole cat was then pulled
up and eaten. None of the other baits was located during the
next two days, and the ravens left.

I later also buried three lamb haunches under light snow
next to where the birds were picking at some almost bare
deer ribs. None of these baits was retrieved in the two weeks
that they were available either. These results do not exclude
the possibility that ravens can smell, but they indicate that
in the winter food that is out of sight is probably not located

by scent. Unless they already know where it is so that they can dig there, the bait is unavailable unless dug up by other animals.

JANUARY 5. It is near zero and snowing hard again this morning, but I'm as close to heaven as I think I'm ever going to get. The balsam fir branches are bent with glistening powdery snow. All sounds are muffled, except for those of forty or so ravens. The yelling and bickering in accompaniment to the constant rhythm of the pounding of their bills on the solidly frozen meat is music to my ears. The only "window" from my cabin this morning is through the 135-millimeter lens of my 35-millimeter camera. The sheep carcass thirty feet from the cabin is almost full frame, and on and surrounding it are the ravens.

I see the snowflakes on their glistening black backs. It is a beautiful sight. I am finally *close* to them, which until now had not seemed possible. The birds are totally at ease. After feeding, some roll on their backs in the snow like happy dogs or lie breast down, fluttering and kicking snow. Some slide on their breasts, pushing themselves forward with their legs. They are snow bathing, something I've seen no other bird do. It looks like young kids romping in new snow, and I'm sure they do it for the fun of it.

I start the wood stove, just to see what will happen when they see smoke. The fire starts—plenty of smoke—and absolutely nothing happens! I detect no change of any kind in the ravens' behavior. So now I cannot only keep warm in the daytime—an extremely rare luxury for me so far—I'm also able to fry pork chops.

It took the birds less than two days to reduce the sheep carcass with its rock-hard frozen meat to a cleanly picked skeleton. I would not have believed it if I had not seen it. Had the meat been soft, they would have undoubtedly finished the job even faster.

Ravens can be very aggressive. Frank C. Craighead, Jr., writes in *Track of the Grizzly:* "Not far from where the grizzlies were feeding, a moving black spot in the whiteness of the snow caught our attention. At closer range we could make out four ravens, one flapping helplessly in the snow

while the others pecked at him viciously. He appeared unable
to fly. The other three, I thought, were surely hammering
him to death. With our approach they flew off, although
reluctantly."

My birds were relatively peaceful even though there was
a good deal of meat to defend. Aggressive encounters (which
I identified as a result of crowding) at choice feeding spots
were restricted to a quick pulling of a wing or tail feather
and only seldom to a chase. Fights seemed to be a matter of
density. When there were only five birds at the carcass,
there was rarely an aggressive encounter; with ten birds,
there were three encounters a minute. With twenty, they
escalated to ten per minute. What a contrast to the blue jays
I saw earlier, which did not tolerate another within some
fifty yards of their bait. And what a contrast with crows,
which showed no aggressive interactions at all.

A COW

How DO I GET enough meat to entice hungry ravens? I drive around and visit dairy farms, and sooner or later I find something, usually a calf.

I walked into Bernie Gaudette's barn in Hinesburg, Vermont, where they had just finished the morning milking. No, he didn't have any calves that died lately. But as an afterthought he pointed to a stall where a huge Holstein cow was lying with her tongue hanging out. Thirteen to fourteen hundred pounds of meat! How could I possibly get such a beast out of its stall to the top of the hill, two hundred miles away and under deep snow to boot? "Thanks, I don't think I can handle one of *those*."

Ten miles farther down the road, however, I had a stern talk with myself. "Do you or don't you want a lot of meat?" I do. "Then don't make excuses. If you are serious, you'll *find* a way." So I turned around and told Bernie, "I'll take her off your hands tomorrow."

You don't learn to dissect a cow in Biology II, but I was making good progress with my sharpened jackknife, as Bernie looked on amazed. Some dozen sixty- to a hundred-pound chunks of cow ought to be a nice raven food bonanza.

The drive to Maine is always a pleasant one. I enjoy it because I anticipate and think about what I will do and how the ravens may respond. It is usually a five-hour trip, but it was somewhat longer hauling a cow packed all around me in the jeep, plus a calf on the hood. Hoofed legs stuck out the windows. At two in the afternoon I got to the foot of my hill, where Dana Eames had come to meet me with his snowmobile. It would take me at least three days to haul up all that meat on foot through the deep snow, but he does it by the time it gets dark, and only charges me twenty dollars.

Now, as I sit in front of the stove anticipating tomorrow, I could not be happier. I've actually got a calf carcass and all the meat from a very large cow up here with me. The only question now is, how to use all this meat most effectively. For the time being my plan is to put it at the old site next to the sheep skeleton that is already familiar to the ravens. They fear the *site* more than the carcass, right? If recruitment is a selfish-herd phenomenon, then they should not recruit because there is little risk to dilute. They already know this site is safe. On the other hand, if recruitment is for sharing the wealth, then the superbonanza should result in more recruitment than ever before. This is my most generous meat offering so far, and I expect unprecedented recruitment. A raven who recruits here will not sacrifice much meat. It can only reap long-term benefits if the others it shares with act similarly in future encounters.

JANUARY 17, 1986. What a beautiful dawn! The red, orange, yellow, fading into blue in the east, contrast brilliantly with the black forest and white snow. It is not even light yet, and a chickadee is already picking at what looks to me like a very bare skeleton of the sheep that served as a meal for at least forty ravens the last time. There is no sign of any raven this morning.

I watch for a few hours and then leave to comb the woods

in the vicinity to see if there are tracks in the snow where
the birds might be retrieving bait they cached previously.
The woods are deserted. After four hours of heavy slogging
I do not find a single track where a raven has landed on the
fresh snow from three days ago. The ravens have not been
back to feed from their caches. Instead, coyotes have been
making the rounds. Following their tracks I find seven of the
ravens' caches that they have dug up. Why do the birds make
caches and then not come back to retrieve the meat?

JANUARY 18. At 8:30 A.M. a blue jay and several chickadees
are feeding on the cow. They are fairly quiet, especially the
blue jay, which does not make a peep all morning.

9:15. A raven has finally found the cow! The great bird's
wingbeats are swooshing so loudly through the air that I can
hear them plainly inside the cabin. It swoops over the cow
three times in five minutes, making a total of forty-five very
deep, rasping, drawn-out quorks. No yells. I hear the quorks
grow faint in the distance. The bird is broadcasting all right,
but not with yells. A half hour later I see two ravens, and
one of them gives a rapid series of high-pitched short calls in
flight. Both birds fly close over the bait, then leave. For the
next five hours or so, I see or hear one or both nine more
times. No birds at all show up after this. But I expect many
tomorrow at dawn.

Near dusk I treat myself to a walk outside in the warm
evening sunshine. And it *is* warm. Amazingly, temperatures
have soared to near 40°F. The surface of the snow is soft and
melting. I smell a wonderful smell. I don't know what it is,
but I associate it with early spring.

The outlines of my tracks from yesterday are dissolving,
and in my footsteps are thousands of tiny black specks. They
are Collembola—the snow fleas! On the first warm day after
subzero temperatures they have magically appeared by the
millions on the top of the three- to four-foot-deep snow! They
must have hopped down from the trees; they could not have
worked up through the thick snow. Is *this* the invisible food
supply that the kinglets eat when they pick at what to me
looks like nothing on the branches?

JANUARY 19. Dawn. No ravens. Daylight. Still no ravens.
Later one comes by *alone*. It returns six more times today.
At least that is how many times I hear one. Near noon, for
the first time, one gives the knocking sounds. What do they
mean? Why are there no yells? During the afternoon all is
silent.

JANUARY 20. At 7:06 A.M., it is barely light. The clouds are
still thick because it rained all night. A raven arrives.
"Blacker was he than blackest jet / Flew low in the rain, and
his feathers not wet," Samuel Taylor Coleridge once described the bird. An adult. Yearling juveniles have dull feathers, and the wings and tail feathers are brownish.

At 9:10 a single raven makes contact with the bait and
stays in the vicinity until 1:03, totally silent all the while.
During these three hours and fifty-three minutes, it makes
ten trips to the bait, feeding each time and flying off with a
bill full of meat. After its last return trip, it did not feed at
all, merely sat in the birch tree above the bait and preened
its feathers. When it finally flew off, I surmised it would not
be back today, and it wasn't. Well, if it brings in others tomorrow morning, I will know it has recruited to share, not
because of the selfish-herd hypothesis, since it obviously considers this meat safe to feed from. It is thoroughly at home
on this bait. No fear whatsoever.

All day yesterday and today the chickadees, two blue jays,
and a red-breasted nuthatch continue to work. You would
think the tiny nuthatch wanted to eat the whole cow, the
fierce way it defends the carcass against any chickadee that
comes near. The raven pays no attention to any of the little
birds. Once, when the raven had flown off, I rushed out and
put my stuffed raven five yards from the cow. When the live
raven came back, it nonchalantly preened in the tree above
the dummy for five minutes. Then it flew down, looked the
thing over methodically and flew back up to resume preening.
This raven always looks sleek and shiny. Its mouth lining
appears to be dark. Therefore it is an adult.

JANUARY 21. Dawn. Most of the snow has melted in the
warm rain. The steady patter on the cabin roof is hypnotiz-

ing, but I'm awakened by a loud croaking—the ravens are back. I see two fly by the window. But not one comes to feed all day today! Maybe they have food elsewhere. Better even than a *cow?*

Six days have passed. My patience is wearing thin. Besides, I cannot stay any longer. Duty calls back home. It seems to have been a week of wasted time. I am disturbed that my predictions have once again been so thoroughly shattered. What in blazes is going on?

The uneaten cow that has been lying up on the hill since January 16 feeds my imagination all week long while I am away. Huge numbers of ravens may be feeding on it, or only two. If only two, it might mean that the birds do *not* really share. On the other hand, if many have come, then recruitment occurred *after* the bait was considered safe, and this would argue for the sharing-herd and against the selfish-herd hypothesis. To drive four hundred miles to find out what is happening is a mere trifle after all the investment I have already put into the cow experiment. Not to go now is like dropping out of a hundred-mile race after running ninety-five miles.

JANUARY 24. As usual, it is dark when I get to camp. I walk up under a nearly full moon on snow that is for once solidly frozen and easy to walk on. Temperatures dip to −15°F. Although I build a fire in the small wood stove, the cabin does not get warm enough for me to keep my hands outside the covers to hold a book. I try sitting next to the stove to read there, but this is not possible, either. After a while I have only one thought—to pull myself deep into the sleeping bag and hibernate until tomorrow.

JANUARY 25. The dawn is eerily quiet, and the whole morning remains that way. Not one raven comes by! After eight days the cow shows hardly a sign of being picked at. Even the fat lies fully intact with only a few small scratchy marks on it.

I might have expected almost anything except *this.* Immediately my mind tries to invent an explanation. Is it that now, in the middle of the winter, the birds have gone to the coast or migrated? No. I check the small town dump only five miles away as the raven flies, and about twenty ravens are there, idly sitting on the trees and on the ground when I drive out to investigate.

The fact remains, two birds found the meat, and at least one fed avidly. Had the lack of an aggregation something to do with the long and continuous knocking I heard last week after the two birds came and neither fed? Maybe the bait was too large and frightening for the *second* bird, who convinced the first of the same thing.

Poring back over my data again, I seem to see a pattern, while animals that should be familiar to them, such as deer or rabbits, were visited with very little hesitation, black-and-white Holstein calves were not investigated for at least two days. Maybe a huge cow such as this, all cut into pieces, was too much to overcome their shyness. But no, that can't be right—one or two *did* feed!

FEBRUARY 1. In the last week the meat has again hardly been touched. The tracks in the snow show that only one or two ravens have been feeding. The coyotes have kept a respectful distance, circling in the woods nearby. I am at a total loss as to why this huge meat pile has not brought in more ravens, because it is now very clear that the birds are not scared by it. Do they again (still?) have better feed elsewhere?

FEBRUARY 2. The night was very cold, and I shivered and had difficulty sleeping. The cold was also letting itself be heard; trees cracked like rifle shots from the frost on an otherwise silent night.

The morning is beautiful. The sky is clear blue, and the chickadees are singing. I, too, feel a change. The ravens should very soon be getting ready to nest. A second winter with them is almost done. But I am even more puzzled than I was before—and two years ago I thought I was going to watch them for two to three weekends and have the answer! The more puzzling it becomes, the more I am drawn in. There can be no stopping now.

TWO RAVENS AND A CROW

SQUABBLES AT A BAIT

BLUE JAY IN A SNOWSTORM
WAITING FOR THE RAVENS TO FINISH

BODY LANGUAGE AT THE FEEDING FRENZY:
THE BOLD, THE TIMID, THE SHOWOFFS

FIG. 11.—A CRITICAL POSITION.

THE LONER IN A NEUTRAL POSITION

CHOKING, FUZZY-HEADED DISPLAY DURING COURTING

**DOING THE JUMPING-JACK:
IS THIS A FEAST
OR A FEARSOME BEAST?**

NEUTRAL POSE

EARED AND FUZZY-HEADED DISPLAYS

RAVEN IN NEUTRAL POSE CALLING

A DOMINANT (RIGHT) DOES NOT DISPLAY,
BUT THE LOWER-RANKER RESPONDS APPROPRIATELY

DOMINANCE AND SUBMISSION

FEMALE (RIGHT) COURTS DOMINANT MALE

A STRUTTING MALE

MACHO MALE DISPLAYING
BOTH TO ASSERT DOMINANCE AND TO COURT

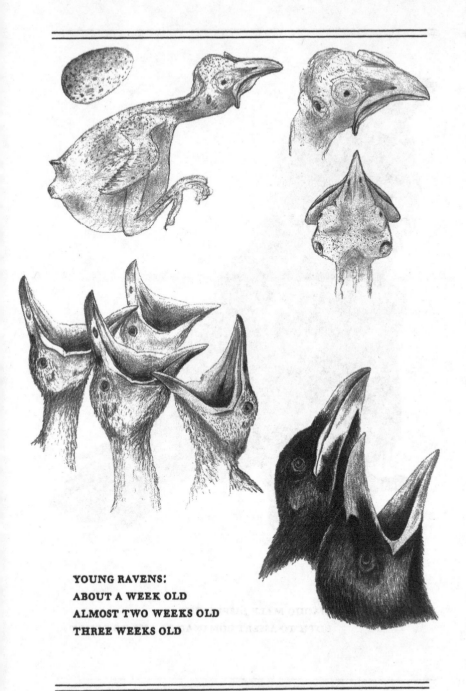

YOUNG RAVENS:
ABOUT A WEEK OLD
ALMOST TWO WEEKS OLD
THREE WEEKS OLD

FLYING OFF TO CACHE MEAT

ROCKING AND ROLLING

A PAIR

Note: A month later when I came back, there was not a scrap of meat left from that cow. Raven feces were splattered everywhere, and the snow was patted down into a solid mat by thousands of fresh footprints. I guessed that at least several dozen birds had recently been there. There may have been well over one hundred. All have left except one bird. This loner remained picking at the well-cleaned bones.

THE LONER

And the Raven, never flitting, still is sitting,
still is sitting
On the pallid bust of Pallas just above my
chamber door;
And his eyes have all the seeming of a
demon's that is dreaming,
And the lamp-light o'er him streaming
throws his shadow on the floor. . . .

—EDGAR ALLAN POE,
The Raven

THE LONER TRIED to sustain himself at the almost clean-picked bones that a crowd of birds had left. For some reason he (I'm guessing it was a male because of its large bill) stayed when the rest went on to find food elsewhere. He stayed for the immediate scraps. But in time he got less and less food, and he grew progressively weaker until he no longer had the option of flying off to search elsewhere or of following others.

The first time I saw him (he could be recognized by some damaged tail feathers) he walked almost in front of my nose next to the window by the cabin. That surprised both of us. He flew off, but rather weakly. In the afternoon he was back, and he made high-pitched trills in a small tree just above the new carcass I had brought. After making forty loud trills, he made beak-snapping noises, soft moans and groans, and croaks. Then he hopped down and cautiously walked toward the bait. To get to the bait he had to walk through a little depression where the window of the cabin and I were not visible to him, and he took twelve tries before he got up enough courage to walk all the way through the blind spot to the bait. He grabbed a rind of suet and then flew off.

A week later he was still there alone (I had removed the meat in the meantime), but he was noticeably weaker. This time I saw him hopping away into the forest when I came up to the bare bones, and he took off as soon as I came into sight. A raven *hopping* away? Strange! It caught my attention, and I pursued him on snowshoes. He hopped into the lower branches of a spruce. I climbed after him, and this appeared to agitate him; he pecked violently at his perch. I persisted in climbing after him, and when we got to near the top of the spruce he flew off weakly down the valley.

I did not expect to see him again. But he was back picking at the same bones the next day. This time I came from another direction, and this time he did not manage to reach a tree before I overtook him in a spirited footrace.

I put him into a cardboard box and closed the cover—after he had eaten butter out of my hand (while I was still holding him when I brought him back to Kaflunk) and after he had also bitten my hand and slashed my cheek. He was shivering when I caught him. With cold or fear?

He might need close attention to nurse back to health, so I took him to Vermont and put him in a makeshift wire cage, which I set onto the kitchen counter.

Released into his cage after a six-hour incarceration in a cardboard box, he did not act excited. He looked around and nonchalantly hopped onto a perch. He did not look weak at all; possibly he was already a little stronger after his first real meal. I offered him a chicken leg. He hesitated, then walked along his perch and took it right out of my hand, hopped to the floor of the cage, and proceeded to tear pieces off and eat. When he was finished, he hopped back up on his perch. I held out a french fry. He walked along his perch to the edge of the wire and took it from my hand and ate it as if he had always known and loved french fries. He also ate a piece of cheese, but he didn't budge when I offered him pieces of orange. How does he know cheese and french fries are good to eat, and orange is not?

All the while, my young son Stuart is chattering, watching the raven eat. The raven totally ignores both Stuart and the cat sitting near the cage, but he keeps a wary eye on me. I am totally amazed at this bird! Ignoring the cat? Is he so interested in *me* because he already knows I am his new meal ticket?

He exudes total control, total ease, right from the start. Not a hint of panic! A bird hand-reared from a nestling could not be more nonchalant. In an hour he shakes himself, rubs his beak on his perch and preens, turning his back to me while I read to Stuart. Right now he is no more than six feet away from me. He yawns, half-closes his eyes as if trying to sleep, rubs his bill some more, nonchalantly picks at his perch, hops down for another snack. All the while he gazes here and there in a way I am tempted to call "thoughtfully." Now he preens again, fluffs himself out, shakes, turns, and idly studies the floor of his cage.

Can this really be a wild bird—the same kind I saw today on the ice on the Androscoggin River, flying up from more than a half mile away after I stopped the car to study the dark spot in the distance? Can it be the same species as the bird that flees in haste from a carcass after I drop a spoon in the cabin, whose bill opens in fright as it approaches a calf carcass it has studied for three full days?

Here is a species of bird well-known for its extreme shyness toward anything "new." Oskar and Magdalena Heinroth, the famous German ornithologists at the Berlin Zoo, in their 1926 book *Die Vögel Mitteleuropas* wrote how they marveled at the way their two tame ravens banged themselves about in their cage for many hours "until totally exhausted" after someone hoisted a flag some 300 feet distant. What could be more "new" than suddenly being in a wire cage on my kitchen table with a boy, a man, and a cat nearby? I was tempted to conclude that my raven was somebody's pet. But just hours before he had tried desperately to escape as soon as he glimpsed me. (I later captured another raven at a dump who also seemed tame within hours. This bird declined rapidly in physical condition, and when it became obvious to me after a month that he would not live, I killed him to end his suffering; an autopsy revealed lead pellets. Other healthy ravens released into a darkened room were also unafraid enough to feed from my hand within minutes. Clearly the Lone One's tameness in capitivity is not unique.)

I had already concluded that ravens may stay away from a bait for days and weeks if they see an incredibly slight disturbance near it. And now, when *everything* is suddenly new, this bird acts as if *nothing* is out of the ordinary! I do not know how they perceive the world. I can only guess that they see it not as an absolute, but as departures from the accepted. When *everything* is different, then comparisons cease, and almost anything can be accepted. And come to think of it, isn't that how humans perceive the world as well?

There are several other things about this particular raven that amaze me. He seems alert. He is far from my image of an animal starving to death. In his cage he looks perfectly hale and hearty. Although, perhaps from starvation, his flight muscles have atrophied, he can hop quite well—as fast as I can run in the snow. I doubt that his wings are physically injured because he shows no sign of mechanical displacement or awkwardness. He can fly, but only the way a person runs whose legs have given out after having "hit the wall."

His curiosity is palpable. He inspects everything I hold out to him. A piece of crust with peanut butter? He comes over, tastes it, drops it, cocks his head to watch it hit the ground! Would *I* be so curious after I'd been starving and caged?

Mostly, it is his calmness that amazes me. He acts as if he had no fears in the world. He does turn his head, but jauntily. He does not once fly against the wire. He hops from one perch to the next when he wants to get down or up. He makes me feel good.

The next day after his capture the raven perches quietly in his cage, but he doesn't miss a thing. His brown eyes roll and snap in this direction and that—up to a fly crawling on the ceiling, down to the cat walking by, over to me making

supper. Every once in a while I stop what I'm doing to talk
to him. He seems to loosen up even more, and responds by
making soft smacking noises.

There are steak scraps, a dead chipmunk (fresh out of my
freezer), an owl feather, and bread on the floor of his cage,
just a foot below his perch. Occasionally he looks down,
studying them. When I hold something up—chocolate ice
cream on a spoon, a piece of bread, a lump of peanut butter—
he walks sideways on his perch and cautiously tastes it. He
takes tiny tidbits into the tip of his beak, works them back
and forth, then swallows or spits them out. He swallows ice
cream, blueberries, fat, and french fries; he spits out bread
(which he ate yesterday), catfood tuna, and orange. He is
getting more fussy in his tastes, but he is enthusiastic about
the chipmunk. He hacks at it, tears apart the hindquarters
bit by bit. It slips from his grasp and drops. Strangely, he
does not retrieve it from the floor, but takes it from me when
I hand it to him. He even comes over to take the owl feather
from my hand, plays with it for a few seconds, and drops it.
And he loves snow! He drinks water out of a spoon that I
hold out to him, but if it is a spoon full of snow, he comes
much more eagerly and eats one mouthful after the next.
Chunks of ice are swallowed whole.

By late April my former kitchen companion has been long
housed outside, in an aviary. He flies over to within about
five feet of me when I enter his roomy cage and offer him
food. Out here he does not take food from my hand, but he
approaches to pick it off the branch in front of me. He flies
vigorously around the cage, and he is silent at all times.

He gets no more ice cream, but he loves dead animals. He
approaches a dead pigeon immediately, without any hesitation. He eventually ate a whole gray squirrel but not without
difficulty; he could not penetrate the skin. His solution:
"skinning" it through the mouth. The end result was a clean
squirrel skin with the fur on the *inside*.

He has been given a clean bill of health. His feces spun
in a sucrose solution on the lab centrifuge showed no para-
site eggs floating to the top. His blood has been examined
under the microscope by a malaria specialist. No blood para-

sites. So it is probable that his problem really was starvation, as I had originally guessed.

His behavior is still fascinating to me, because it helps me interpret wild birds' actions. Take feather posture. When he is afraid of me and backs off, he sleeks back the feathers on his head. (Strange, those birds who are subordinate to other ravens in the field, who back off at the bait, have elevated feathers and "fluffy" heads.) But now in my presence he is often enough at ease to fluff out his body feathers and even to shake himself.

Today, April 21, 1986, I capture him with an insect net, wrap him in a jacket, and put on his leg the metal ring from the U.S. Fish and Wildlife Service and also a green plastic wraparound ring from National Band & Tag Company. The band is about half an inch wide and wraps once about itself. I do not think there is any way to get it off. But he is frantic immediately and keeps biting and pecking at the plastic band. I go away, and two hours later when I come back, his right leg is bloody—scales are off and bone is exposed—and the bandette is off! He tolerates the aluminum ring. I had originally thought it would be sufficient to color-code wild birds with plastic colored rings, because you can see the colors with a spotting scope. How much effort I would have wasted if I had not had the opportunity to test the plan, which seemed so surefire and in no apparent need of testing! Now I know I cannot catch wild birds and band them with plastic tags. (I have banded nestlings with the same bands, and they ignored them. However, many parent birds eject bands from the nest, along with the attached nestlings.)

My loner is soon "singing" every day in his cage. A singing raven warbles and squawks and yells and trills and goes through a large verbal repertoire.

I released the raven on July 10, 1986. He then stopped singing entirely, but he stayed near the house, being quite sociable with the ravens I then had in both of the other aviaries (see next chapter). On July 20, I released other tame ravens as well, and he struck up mutual friendships with them, especially with one. The two left together, and on July 28 he dropped in on some picnickers at Little River State

Park, in Waterbury, Vermont. He must still have been quite
tame, because they were able to read the number of his
aluminum leg band, which was reported to me months later
by the U.S. Fish and Wildlife Service. Apparently, like other
ravens, he has now learned what *not* to fear: humans. So if
you ever meet up with a raven who just loves french fries,
Ben & Jerry's chocolate ice cream, and squashed chipmunks,
see if he has an aluminum leg band. If it is U.S. Band No.
706-21301, please give me a ring. I'd love to hear about the
latest adventures of an old good friend.

TAME BIRDS FROM THE NEST

THE ULTIMATE AIM of behavioral research is to understand organisms in their natural environment. This is not always an easy task, especially in shy and free-ranging animals. Compromises are usually necessary, first in one aspect and then in another, until one finally gets a balanced, and more complete, perspective. No one approach can provide all the answers because each has limitations and faults. But one can get an accurate overall perspective by taking many approaches and gleaning the best that each has to offer.

I felt that I could not really understand or even see the raven problem until I understood or had intimacy with the animal. And I knew I could not understand ravens by seeing them only from a distance in the field, anymore than I could understand them by studying one in a cage in my living room, especially since I knew nothing of its past. Did the apparent fear of baits I saw in the field relate to ravens having been trapped, or seeing others trapped? (An idea given to me as the "obvious" reason by another researcher on crows.) If this were the case, then hand-reared birds should show relatively little hesitation in attempting to feed, and they would then be a neat way to test the selfish-herd hypothesis; if the bait is considered safe there should be no recruitment.

There were other reasons why I wanted to work with hand-reared ravens. From the example of the tame crows I have kept since I was a small boy, I expected tame ravens to follow me. This gave me a bright idea—perhaps I could keep a group in a large outdoor aviary, release one, and lead it to

a juicy deer carcass. Would it then recruit its fellows? And
if so, how? Furthermore, I could test whether or not relatives
are preferentially recruited by having one aviary full of
brothers and sisters, the other containing only strangers.

Aside from having specific questions to which tame birds
might give an angle not visible in the field, I felt that the
real value would be unanticipated insights, which one often
gets when exploring previously untraveled ground. There
was also the personal, atavistic reason of wanting to know
another "blood"—another, though different, warm-blooded
animal like oneself—which could come only by getting to
know individuals. Most of the other research necessarily involves treating the population as an entity, gathering data on
it, and then analyzing the organism at arm's length through
bar graphs, statistics, and figures with X and Y axes. General
results emerge in terms of averages and differences about a
mean, but no matter how thorough, such information can
never reveal one very important attribute: *individual* differences. These differences, perhaps even "personality," are
more than a variation about the mean. And they are as much
a real part of at least some animals as are the collective
shapes of their beaks or the collective workings of their guts.
For the most part, however, individual differences are considered a bothersome variable that tends to be minimized because it gets in the way of "consistent" results or "averages."

By crossing the same territory from different directions simultaneously in order to find the pulse, one may be accused
of being "unfocused." But in biology the range of possibilities
is often large, and unless one is very lucky, one has to make
at least a superficial scan of numerous possibilities before one
can have any confidence of bearing down on something solid,
alive, and relevant.

With some of these unspoken thoughts in the back of my
mind, I procured the necessary state and federal permits and
took the young ravens from two nests in May 1986 for the
ostensible purpose of "experimentally testing recruitment."

When they were young, I kept them in hay-filled buckets,
which I transported back and forth from my home to my
office in Burlington, Vermont. They needed to be fed about
every hour, and they voraciously ate all the roadkills that I

picked up and chopped for them. I supplemented this diet with canned dog and cat chow, raw eggs and cottage cheese, insects when I could get them, and other varied fare in small amounts. In general, they quickly tired of any one item of food, regurgitating it if they received it too many times in a row.

Near the end of May, when they were ready to leave the nest, I put them into the large aviaries I had built for them next to the house. They woke me every morning near sunrise by their penetrating loud calls for food.

One clutch, the Dixfielders, consisted of two young, Theo and Thor (later alias Snoopy); the other of three, the Welders (Ralph, Ro, and Rave). When I got them, the still naked and sightless young of both clutches begged in response to any disturbance by stretching their necks straight up and gaping, without regard to the direction or sources of the disturbance. At least before they fledged, I could not distinguish the individuals, and it is reasonable to assume that I gave them all equal treatment. (After they fledged, I could easily distinguish them on the basis of behavior as well as physical appearance, such as bill shape and size.)

The two clutches of birds were kept in separate aviaries, and many of the detailed observations were made twice a day (for one half to two hours) through a picture window of my house onto which one of the large aviaries abutted.

Different personalities emerged in the Welders even before the birds fledged. For example, in mid-May, when the young were all of equal size and fully feathered out, Ro and Rave rested peacefully in the nest, while Ralph engaged in an almost continuous round of stretching, wing flapping, and picking at the branches of their reconstituted raven nest. He spent much time looking attentively all around. As soon as he left the nest, he developed the habit of jumping on my, and other people's, arms and shoulders at every opportunity, primarily it seemed to peck at buttons and earlobes, and to yank on any loose clothing. However, he did not allow me to *touch* him. None of the other five birds *ever* landed on anyone even once.

Ralph was the most adventurous of the three, the only one who habitually peered into the windows of my house and tried to enter it. He was the first to use a new roost (which they later all used), take a bath in the pan provided, catch an insect, scold a turtle put into the aviary, peck at a pot, approach a stranger, use a new perching tree, and kill a small frog. Most conspicuously, he was *always* the first to investigate any new food placed on the ground; consequently, he always had followers. I at no time observed a single action initiated by Ro or Rave. Ralph was the "initiator" or leader, and Ro and Rave the followers, who went down to feed only after Ralph had been there first. For example, I spread a handful of cooked rice on the ground in the aviary, which all three birds ignored for a day. On the second day, I picked up some of the rice and held it up to Ralph. As chief raven keeper, I was apparently top bird, and so I was the initiator for Ralph, who eagerly ate the rice from my hand. He then hopped down to pick at the rice on the ground, and within another minute both Ro and Rave had joined him. This pattern of behavior was followed with small animal carcasses or any new food that I provided.

Thor of the Dixfielders also trusted me, and I could serve as leader for these ravens. Neither Thor nor Theo picked at a run-over bird for the one day that it lay on the floor of the aviary. I then held the bird up, and Thor, the leader as usual, immediately flew over to me and picked at it in my hand. He continued to pick at it when I returned it to the ground. Theo immediately joined in and fed from it also. Exactly the same behavioral sequence was observed repeatedly in these birds that I had observed again and again with Ralph of the Welders.

Shortly before he fledged, I described Ro as "the slowest, dullest, and least adventurous" of the three in his clutch. Three months later, after he had molted to adult garb (except for wing and tail feathers), that assessment had not changed. Unlike Ralph, he showed no overt curiosity about novel objects or novel food placed in the aviary. None of the other birds associated with him, and when the birds were later released, none of the other birds followed him.

Others have recorded the great variation in behavior between individual captive ravens. Eberhard Gwinner in Germany observed great differences in play behavior in young birds and in nest building. In one pair, the male built most of the rough nest, and the female provided the lining. In another, it was the reverse. In a third pair, both worked equally throughout, and in still another pair, the male took almost no part in the nest building.

The Dixfielders also had strikingly different personalities. Three months after fledging, both were tame and fed from my hand, but Thor was the only raven of either clutch who allowed me to *touch* him. Indeed, he sat still in obvious relish, closed his eyes and made soft, cooing, nasal comfort sounds while I threaded my fingers at will through the feathers of his head, neck, back, breast, and belly for as long as I liked. He did not, however, ever perch on me.

Thor was the leader of the Dixfielders, and like Ralph of the Welders, the boldest, liveliest, and most inquisitive of the brood. He was caged with his nestmate, Theo, and later with Ro. Thor was always the first to investigate any strange object placed in the aviary, including peanuts, grain, cereal, various insects, dead birds and mammals, dishes, utensils, flowers, earthworms, reptiles, amphibians, fruits, brightly colored rings, blocks, toys, silverware. Theo clearly was also *interested* in the objects, but she observed them only from a high perch until Thor contacted them. Ro, in contrast, appeared totally uninterested, as he had when he had been with his own clutch, barely bowing his head, except toward some of the larger conspicuous objects (a dead raccoon, etc.). If Thor left an object without feeding on it, Theo never bothered to go down to investigate it. She (I assumed she was a female because she was smaller than Thor) was interested in objects only as long as Thor showed interest in them. Whenever Thor started to feed, Theo joined in or tried to join in within a minute or two. Alternatively, Theo followed Thor around and tried to extract food directly out of his bill, even though there was food of the same kind readily available in the cage. If the food was new, she was interested only in those individual morsels that Thor was feeding from, even though she had to struggle to get a few bites. If I held a leaf up to Thor, she tried to take it from him. If he ate snow, she ate it, too, but only out of his bill.

A furry animal the size of a gray squirrel would, the first

time presented, not be touched by Thor until he had examined it for some minutes or even hours, first from a perch,
then from closer on the ground. He would advance cautiously
and peck, retreat, peck again, and so on for several minutes,
exactly like the wild birds I had observed in Maine. And
Theo would not approach the squirrel until after Thor had
fed from it, some time later. Once, Ro showed no caution. He
approached directly, without hesitation. Within seconds he
had landed directly on top of the squirrel like a hawk and
started pulling out the entrails. But his feed was short. Both
Thor and Theo were watching, and in less than a minute
Thor jumped down, grabbed Ro by the tail and slowly
dragged him away, then jumped on him and pecked him. Ro
retreated to his night roosting perch. Thor now fed alone, but
Theo soon tried to join in. She, too, was shooed back, but not
attacked. For twenty-five minutes Thor had the squirrel all
to himself, while Theo kept trying to intrude. Before I
stopped keeping count, she had been repulsed twenty-one
times in succession.

After half an hour Thor started to take breaks to cache
squirrel meat. His objections to Theo's interference became
more and more perfunctory, and soon both birds were feeding together side by side. Ro did not intrude but continued
to watch from his perch.

After both Thor and Theo were satiated, I provided a run-
over bird. Ro, seeing uncontested meat, immediately came
off his perch and started to feed avidly, but Thor and Theo
also came over. Ro was aggressive toward both, and now,
fully fed, they hastily jumped aside. But I thought it was less
submission than just not wanting the bother of dealing with
him.

Later that same morning I gave the birds a freshly cooked
ear of sweet corn. As usual, Thor, the leader, was the first to
investigate this strange and, to him, fearful object, to peck
it, jump, peck again, etc., until eventually he began feeding.
Theo, as usual observing from a distance, tried to join in
immediately after Thor took his first bites. But Thor vigorously chased Theo fifteen times in succession. So Theo ate
left-over squirrel. Ro stayed on his perch.

I wanted to glut Thor and provided three more ears of

corn. Now Ro also got a chance to feed on an ear. When Theo or Thor came near him, he repelled them every time! Apparently because of hunger he valued the corn more, or they, because of satiation, valued it less.

These observations from only one day show that for ravens feeding is a very social activity. Thor, the leader who examined all strange objects, was clearly the dominant bird. He could take from the others what he wanted and when he wanted it. But he had no fetish about wanting to be a "winner"; like an apparently submissive individual, he readily stepped aside from both birds to avoid a fight when there was little to be won.

Only Ralph and Thor, the two dominants, showed conspicuous bravery. But initially even these birds showed fear at anything new in their respective cages. The fear was most apparent with large objects—a dead woodchuck, a chair, some stangers—when the birds were still young. They responded by flying wildly about in the aviary, banging against the screening. Less threatening objects—a kettle or a dead squirrel—might cause one bird to retreat to its roost, while another observed the thing from a closer perch. Eventually the leader hopped down to the ground to check it out. Exactly as I had observed in the field with wild ravens at carcasses, the leader edged up to the object gradually, jumped back, edged up again, until finally he came close enough to strike it with his bill. After several approaches and pecks, he picked up the object (if it was small enough) and dropped it again quickly in the same motion as he jumped; the object was thrown into the air and the bird made loud, angry, rasping calls. Usually after several pecks and throws, the bird investigated the thing at close range, and either fed on it or completely ignored it from then on.

I observed this behavior when I presented for the first time a spoon, a fuzzy caterpillar, a small frog, a sprig of goldenrod, a handful of green vines, a handful of Cheerios cereal, a pork chop bone, and, of course, any furry or feathered animal. The bone, Cheerios, flower, caterpillar, and vines did not cause alarm when encountered a second time, but the live frog and dead birds and mammals continued to bring out the same response, though in milder form, on several more encounters. The inanimate and/or inedible objects were later totally ignored. Most insects—a dragonfly, butterflies, large beetles, *smooth* caterpillars, grasshoppers, wasps, and flies— were approached eagerly and eaten without hesitation, even when encountered for the first time. I conclude that the exaggerated bait shyness (neophobia) that I had seen in the field is innate. The ravens do not have to learn to avoid carcasses; they already have a healthy respect for them, and the "jumping-jack" maneuvers they use to investigate new situations cannot be an attempt to snap traps shut so that feeding can begin, as some people have suggested. The ravens fear something with which they have had a bad experience less than they do something that is new to them. My studies also showed that there are many individual idiosyncrasies of behavior, and that ravens have dominance hierarchies within a group that likely link high status with courage or daring.

I learned a great deal from the captured ravens that would assume significance later, but one experiment I had wanted to do could not be done. I had expected the ravens to stay after I released them. But unlike crows I have had in the past, these birds have a strong tendency to wander. Within days they dispersed and I had difficulty recapturing three of them. Since I was their parent for most of their lives, I assume that raven parent-offspring bonds are easily dissolved. (Later field studies showed the same.) To me this was an indication that if I wanted to look for the mechanism and reasons for recruitment among ravens, I should probably abandon the usual explanation of kin selection.

One more totally unexpected finding was mouth color. In January when Theo and Thor were housed together without rivals, except for Ro, their mouth linings turned from bright pink to black. All other similarly-aged birds retained their pink tongues and pink mouth linings. This is a significant observation that merits more detailed study, because several publications had previously indicated that mouth color is diagnostic of *age*, with juveniles having pink mouths and adults black. What was different about Theo and Thor was that they had no serious rivals, and that they courted each other throughout every day. Many birds change bill color under hormonal influence as they enter breeding conditions, and it is also well known that breeding condition in birds is highly sensitive to psychological stimuli. According to an acquaintance, another raven he reared alone also "prematurely" changed its mouth color from pink to black in the first fall. The well-known dominance structure in ravens deciphered by Eberhard Gwinner could therefore act to physiologically suppress maturation in the individuals who are dominated; that is, dominance status probably bestows breeding privileges by acting through the endocrine system, as is well-known for social insects.

ANOTHER HYPOTHESIS

OCTOBER 20, 1986. My third season of raven-watching will now begin. Today is my first trip out to the study area. The first thing I want to do is find out if there are ravens about, and if so, if they still recruit. I need to see it again, live, to make this as-yet-unbelievable and unexplained wonder seem real so that my energy is recharged to look for the explanations behind it.

As usual, when I first get to Maine, I check with Mike Pratt, the local game warden, on the off-chance that he had his hands on a dead animal. He does. Two moose (a cow and her calf) were found dead ten days ago down near the outlet of the lake. Shot dead and left. Killers on the loose. He gives me directions to the spot in a clearing on a remote logging road where they now lie.

As I drive down the logging road through young alders and around big potholes to look for the moose, I keep an eye out for ravens. But there are none to mark the spot; instead, the stench leads to it. The calf, which a trapper had cut open to take some meat, has already been completely eaten by ravens. Only the skin and the cleanly picked skeleton remain. The cow, in contrast, is bloated but intact except for her eyes, which have been picked out. The white blotches of feces running down all sides of her back show that ravens have used her as a perch. Her front end is dissolving into a huge, undulating mass of maggots, as the blowflies have gained entrance through the head. Clearly the ravens, given a choice, like neither maggots nor rotten meat. But the hindquarters have not yet been penetrated, and I cut open the

146

skin and peel it back to expose the red unspoiled meat. Now that the meat is exposed, will the ravens come?

I leave at three o'clock, two hours before dark, not having seen or heard a single raven in the area for an hour.

OCTOBER 21. When I return to the moose in the morning, there are already five ravens feeding. Here are the clear results of an unplanned experiment showing that carnivores/scavengers (in this case, me, who acted in their place to open the carcass) are necessary for the ravens to compete with bacteria and flies for the same food resources.

It is a rare opportunity to have a moose carcass, or even a part of one, available. By dusk my nephew Charlie, who has come to help, and I have woven a blind out of young fir trees and branches. We drag the moose downwind with the jeep to within fifty feet of our blind, where we will begin our vigil tomorrow morning.

OCTOBER 22. I settle onto the cushions of fir boughs in the blind and within minutes hear ravens in the area. But not until two hours later does the first bird come down to feed. Within two minutes it is joined by four others. After that the number of birds simultaneously on the carcass ranges from five to almost twenty for the rest of the day. (There were so many, and they crowded so closely together on the hindquarters where they were feeding, that it was difficult to make exact counts.)

Birds come and go continuously, and almost none leaves without carrying a huge chunk of red meat crammed into its expandable throat and extruding from its beak. With an average of 7.1 birds there during one nine-minute count, there were thirteen arrivals and thirteen departures, each with a load of meat. The traffic goes on at this rate without pause all day. By the end of the day about one third of the meat has been stripped from the hindquarters. While the ravens are eating their way gradually toward the front of the moose, the maggots have the front all to themselves and are continuing to move toward the back. Ravens have been reported to eat maggots at carcasses, but these certainly do not.

There is no serious fighting among the ravens all day. Occasionally a bird squawks as it is apparently shoved aside by a more dominant bird in the midst of the densest aggregation of feeders. But there is enough space for many, and with ten birds there, I saw only one mild squabble in a ten-minute observation period.

Instead of fighting, I see courting! Often in the trees above the moose there are pairs of birds making soft mewing or cooing noises. (I now know that these need not be adults, because my caged juveniles started courting even by late summer.) Occasionally one of the two fluffs out its feathers, sidles up to the other, and makes bowing motions. The visual part of the displays is the same between different birds, but the sounds they make are often individually distinct. One bird went: "(a snap)-quork," "(snap)-quork;" another: "(a squeak)-snap," "(a squeak)-snap," or "(a snap)-grunt," or "(a snap)-hiccup," and so on.

Some of the squabbles that developed were apparently not just over food. On three occasions I saw a bird displaying itself to another bird in the tree above me, and when a third bird came close, the first attacked it. Three other times I saw extended aerial chases, during which there was much excited high-pitched cawing. On at least six occasions all the birds left suddenly, as if alarmed by something (I was never able to determine what).

Five times I noticed that the bird who first descends to feed and to lead the others in, either yells or makes a series of three knocks in a rather slow sequence, in contrast to the fast "knocking"—the xylophonelike sound with many short calls per second, starting at high pitch, and descending the scale to terminate in about two seconds with an abrupt little "thunk." While making the "knock, knock, knock" sounds, the raven assumes the display posture of body and feathers described for dominant courting birds. While he is doing this, I always see an audience of other birds nearby that seems to be waiting to come down. Why are they waiting? I do not know, but I quickly become convinced that I have been wrong about my interpretation of the knocking. It is probably part of a courting or dominance display. Perhaps I heard it near baits after ravens had left, not because they

meant to warn of danger, but because they were idling about
rather than feeding, and thus had time to display.

Most songbirds do their courting in the spring, so it is easy
to dismiss potential courting behavior now, in late October.
But ravens breed in the winter! The young males may have
to win a female and a suitable territory well before fall in
order to be incubating eggs in March.

The observations at the moose are an unexpected bonanza
for me. I had actually brought my own bait—two hundred
pounds of pig and cow offal, collected at slaughterhouses, and
two hundred pounds of cracked corn. Charlie and I distributed the meat and grain in seven different piles each along
old, untraveled logging roads in the woods. Would there be
differences in behavior at the two different kinds of bait?
Would the birds distribute themselves equally between the
different stations, or would they cluster? How long would it
take them to find the baits? It is a very different experiment,
and therefore might yield unknown and unanticipated results.

The results from the grain piles were simple: No ravens
fed from any of them. Even though in the western United
States ravens can subsist on corn in the winter, these ravens
did not even touch the grain left at the place where the moose
had been. (Was it too hard and dry? My tame ravens loved
fresh corn.) One grain pile was visited by a robin, another
by several ruffed grouse, another by two crows and a deer.

Unexpectedly, the results from the meat piles, which were
interspersed between the grain piles, were very exciting and
complex. Two patterns emerged: There was either a lot of
very fast recruitment, or no recruitment at all! Three of the
meat piles that were visited by one raven never attracted
more than two, and they looked as if they were fighting other
birds off. At the meat pile near my cabin, we saw two ravens
begin to feed at dawn without making a single sound. Then
we saw a third in the vicinity, after which we heard the
harsh grating calls birds make when they fight. The next
day there were still only two birds at that meat pile. On the
other hand, at another, almost identical meat pile no more
than a mile distant, there was a buildup to twenty-five birds
within one day. Two other meat piles each attracted at least
fifteen birds at the same time, again within one day. Clearly

it is not the *size* of the food bonanza that is the critical variable for recruitment. Are identical meat piles *either* defended *or* recruited to? This is preposterous, but it is consistent with the data so far. I don't like such unpredictability, because I want uniformity of results that can be potentially understood. But I can't help being intrigued by strange observations. It means that I have been overlooking something very important, and now I must search for it.

NOVEMBER 4–9. The experiments comparing ravens on meat versus grain piles have been the most exciting yet, but the results are not at all what I had planned. I am seeing something totally unexpected. Vague suspicions and speculations run through my mind, and I barely admit them to myself because of the large implications. I will now repeat the unplanned experiment I did last week in earnest, this time putting out fourteen bait stations of meat only. All baits are within my study area, where I had trouble getting more than two birds to come to the cow. If many ravens now come to any one of the fourteen baits, it could hardly be because that site is special.

It has just snowed and I can now see tracks. I could not have picked a better time to monitor the results of this experiment.

By evening two of the meat piles have already been discovered by ravens, and up to a dozen birds have already gathered at one site by dusk. Shortly after daybreak the next morning, up to thirty ravens arrive at this pile, while twelve other identical meat piles remain unvisited. Clearly it is not only fear of the bait or the site that so often causes baits to remain unvisited. Furthermore, if birds had been searching and arriving at baits independently, they would have had a nearly equal chance of finding any one of the other twelve baits. There is no doubt that the ravens' distribution at baits is clumped. It is so clear I need no statistics to prove it. To do more to convince people would be to imply they are idiots who can't see.

There are other instructive lessons to be gleaned from this experiment. First, four baits that attracted two ravens each have no recruitment at all over four days! Why, then, did *other* baits show increases in numbers almost immediately?

The many baits I had put out this time were along deserted logging roads, and the birds began feeding within two to three hours after discovering the food! During the previous two winters when I put the food near my cabin, I had never seen ravens begin to feed, and to recruit, so rapidly. The birds' delay in beginning to feed had always been very puzzling because a predator could have removed the meat within hours, and such slow responses to such good bait seemed to me maladaptive, to say the least.

Combining these observations with those of my captive birds, I think back to my original hypothesis. Maybe the ravens did fear the cabin, and it took a long time for a leader to emerge who would go down to the bait. However, after the leader acted, the rest were suddenly fearless. But that still does not explain nonrecruitment to baits that are fed from!

I noted earlier that at the cabin there was never recruitment shortly after a bait was discovered. But almost invariably there was recruitment *eventually*, given enough time—days or weeks—and a large enough bait. Now I see it differently. There may be two things going on simultaneously: Under one situation, there is immediate recruitment; under the other situation (at the same time, the same area, and the same kind of bait), there is *no* recruitment. The difference must lie with the birds themselves. Could the nonrecruiters be the "married" pairs that stay year-round, year in and year out, in the same territory—like the Hills Pond pair that nests close to my cabin? My young birds did not stay long in their "home" territory by my house. Perhaps the recruiters are vagabond juveniles and subadults who do not dare go near any bait that a pair holds. But a raven's territory is likely to include many tens of square miles. Wherever a vagrant juvenile finds a carcass, it is likely to be in some adult raven's territory, and if this lone "scout" finds it first, perhaps it recruits very rapidly to overwhelm the territorial defenders who might already be there or who will certainly come quickly. Is *that* why recruitment is so dramatically fast, when it does happen? Of course, a raven pair is not continuously at any one bait. The critical and very beautiful

unplanned experiment I had performed without really knowing it was this: By providing *many* baits in one area, I made
it impossible for the defending pair to be everywhere at once.
The vagabond crowd could move in to one or another bait
unopposed while the pair was occupied defending some other
bait. It is a neat hypothesis consistent with the results. But
like all the others, it is an idle one, unless I can get solid data
to prove that there really are juvenile vagrants and adult
residents, and that the two behave differently at baits. If one
could only *identify* individuals in the field! But with ravens,
I am afraid, that may remain impossible; I have heard about
many fruitless attempts by other researchers to capture and
mark ravens. I will probably have to rely on massive volumes
of observations to try to get insights obliquely.

TERRITORIAL ADULTS AND
WANDERING JUVENILES

MONG THE CROWS and their relatives it is almost universal for adults to form permanent pair bonds. The dominant, generally larger male is responsible for and alone capable of holding a permanent territory. The young are fed by both parents, and they stay with them during a period of dependency that lasts generally from a few weeks to several months after fledging. When they leave the natal territory, the young often associate with other juveniles, and for the common raven, *C. corax;* the Australian raven, *C. coronoides;* the thick-billed raven, *C. capensis;* the carrion crow, *C. corone,* and even the magpie, *Pica pica,* there is in the literature repeated mention of "flocks" of juveniles in search of food.

Each pair of adult ravens that has formed "permanent" pair bonds resides year-round in its territory or domain. The pair sleeps nightly very close to the nest, which is often repaired and then reused year after year. The domain is defended year-round against other ravens, in contrast to the defense against aerial enemies of other species, which varies seasonally and is restricted to the immediate nest area. There is considerable behavior flexibility and perhaps subtlety in territorial defense. The East German raven researcher Johannes Gothe saw a pair escort (rather than attack) a trio of strangers out of its territory by flying on each side of them. Territorial behavior may thus not always be easy to detect, but it is evident indirectly from the spacing between neighboring nests.

Gothe has mapped nearly all of the raven-nesting sites in

the Mecklenburg area of northeastern Germany since ravens
returned to that area in the early 1940s (apparently from
Scandinavia via neighboring Schleswig-Holstein). When raven populations there stabilized in 1955, there were seventy-
three active nest sites. In 700 square miles of prime beech-
wood habitat there were thirty-seven breeding pairs, giving
an average of 18.6 square miles domain per pair. This high
population density was achieved not by a gradual geographical expansion of the species into the area, but by increasing
density in the existing area by several apparent waves of im-
migration. Gothe therefore suggests that raven pairs space
themselves out without defense of specific boundaries when
population density is low, but that they defend ever-more-
defined boundaries as the available habitat becomes saturated.

Recent studies in Britain by M. Marquiss and I. Newton
from the Institute of Terrestrial Ecology in Scotland and by
D. A. Ratcliffe, J. E. Davis, P. E. Davis, P. J. Ewins, and
J. N. Dymond of the British Nature Conservancy support
Gothe's conclusions and show that breeding density is also
very much determined by food supply. The highest breeding
density of ravens in the world seems to be in central Wales,
where, because of a somewhat artificial situation of abundant
sheep carrion, territories are compressed to only 2.0 square
miles. In northern Wales, raven densities have increased
since the 1950s, thanks to increases in the availability of
sheep carcasses, until a breeding density of 3.7 square miles
per pair was achieved. Before and after the increase, the nest
territories were, as in Germany, regularly rather than ran-
domly distributed, again arguing for Gothe's idea of active
spacing. In southern Scotland and northern England and in
central Scotland, better animal-husbandry practices, resulting in fewer sheep carcasses left on the sheepwalk, and re-
forestation, leaving the sheepwalk less open, have resulted
in a nearly 50 percent decline in raven populations since
1960. In Virginia, Robert G. Hooper and colleagues from the
U.S. Forest Service found that about eleven square miles supported one raven territory.

Coastlines seem to provide abundant food, and the coast of
Shetland supports a dense population of evenly spaced nests
on cliffs. In England, up to fifteen pairs of ravens have been

reported from only seventeen miles of coast near Devon and Cornwall, with a maximum of three pairs along one 1,000-yard stretch of cliff. Here the birds' domains may have been compressed because of localized abundant food, or else the feeding domains were separated from the nesting sites. In general, a 1.2 mile stretch seems to be the nearest distance that ravens will tolerate between nests.

The fledged young initially remain near the nest site, then follow their parents for about three weeks, after which there is apparently mutual estrangement between parents and young. The young voluntarily leave and never return, while new young from other areas may come and take their place. By summer's end when the young start to wander, the parents return to their old sleeping place by the nest. In northern Europe, individual young have been found to move up to 310 miles in the fall and winter of their first year.

The general consensus of most raven researchers is that the juveniles form vagrant "swarms" or flocks. The circumstantial evidence for this is that (1) large groups of ravens have been observed even in the breeding season, and (2) breeding does not occur until the birds are three to four years of age. However, D. T. Holyoak and D. A. Ratcliffe speculate that the swarms are the nonbreeding "surplus" of the population, rather than representing an essential and distinct social phase of the ravens' life strategy.

There is little or no distinction in the literature between pseudo "flocks" of random individuals that happen to gather at food because it's there, and coherent groups or flocks that fly in search of food. The difference is crucial to the point of my study. If the raven crowds at food are the result of nomadic flocks having found that food as a group, then recruitment is a moot point. Gothe cites numerous examples of "swarms" or flocks of ravens that range from about twenty-five to nearly four hundred in number in northern Europe. But no one has determined whether or not these groups are cohesive social "flocks," whether the individuals are indeed juveniles as presumed, or whether the gatherings are random aggregates.

I do not dispute (nor accept) the widely held belief that young ravens form wandering flocks, because no "flock" has

ever been followed to ascertain that assumption. I want only
to distinguish the assumption from the evidence as it regards
specifically *C. corax*, the raven. From a number of perspectives it is eminently plausible that the assumptions are correct, but the gulf between the eminently plausible and the
actual is often immense. Having examined the literature on
the wandering-flock-of-juveniles hypothesis, I believe it is
based more on plausible inference than on empirical evidence. There may indeed be wandering flocks, but not all
gatherings of ravens are necessarily flocks, and not all flocks
necessarily wander, and not all members of flocks are necessarily juveniles.

The main reference for the idea of wandering flocks of juveniles comes from natural-history observations in Europe. The
argument for juvenile ravens joining *Schwärme* (swarms)
is drawn from anecdotal sightings of groups of the birds at
dumps and other food sources.

In biology the term "swarm" presupposes a coherent group
like that of honeybees, with whom the term originated. The
swarm is a social unit that stays together. The word cannot
be applied to a group of bees gathered at a good food source,
no matter how many individuals are involved. The distinction between a swarm and a gathering is therefore crucial to
understanding the phenomenon as applied to ravens. There
is no evidence at all that a raven "swarm" at a food source
is analogous to a swarm of bees. It could simply be individuals aggregating independently.

Rolf Hauri, a pioneer raven worker in Switzerland, berates
Scheven and Schmidt for *assuming* that the sociable "swarms"
consist of juveniles. He warns that really to identify the birds
as juveniles (by details of mouth and plumage color), one
would have to hold them in one's hands. I agree. But his
rationale for them being juveniles is circular: He says they
are juveniles because the adults remain in their territories
while the young leave their parents by late summer, yet
probably do not breed until they are at least three years old.
English translations talk of "flocks" rather than "swarms,"
but flock is not defined.

There is, however, strong evidence that ravens are sometimes sociable and sometimes move en masse. The patterns

or reasons for these mass movements are so far totally unknown and unpredictable, but they are so dramatic that there
is no doubt they exist. Groups of ravens have been observed
in situations seemingly unrelated to food or gathering to
roost. Clayton M. White and Merle Tanner-White at Brigham
Young University in Provo, Utah, describe a gathering of
over 1,000 ravens on April 11, 1982, at midday near Monti-
cello, Utah. Another western biologist, Terry Root, told me
that she once saw "about 2,000 ravens" at around 3:00 P.M.
in 1981, soaring by a ridge of the Sandia Mountains in New
Mexico, an occurrence she told nobody else "because they
wouldn't believe me," perhaps because such sightings are
rarer than UFO reports. It seems doubtful that this many
birds would gather for feeding. Gothe reports having seen
several flocks in transit in a southwesterly direction, apparently from Scandinavia to Germany. An apparently similar
phenomenon was observed near Petoskey in northern Michigan by Kathy Bricker, a biologist and wildlife photographer.
At noon in early December she saw approximately 150 ravens gather, many soaring in pairs, and then descend to
perch in the forest nearby. There was no known food and no
roost. The birds were gone the next day and not seen again.

There are other scattered accounts of groups of ravens apparently on the move together. George T. Austin, in his extensive survey of ravens along the roads of the Mojave
Desert, saw a flock of forty at 6:15 A.M. on February 27,
1968. Five hours later they were gone.

R. Hewson on several occasions saw ravens in Wales appear "from all directions," converging in an aerial display.
On November 25, 1942, he watched three to four pairs soaring at 3:00 P.M., and twenty minutes later there were twenty-
five birds. In Maine, I once observed a crowd of thirteen
ravens—six flying by two's and one single—in early February. Fifteen minutes later, ten more ravens from a known
feeding site joined the highly vocal crowd, which then vanished from my view.

I had the good fortune to witness another mass flight at the
Hahnheide, a nature preserve within twenty-five miles of
Hamburg, West Germany. At about 1:00 P.M. on August
27, 1988, I heard the aerial chase calls of a raven and looking

up saw three birds in a mutual display, during which two birds were usually close together in flight, with the third following within one to two yards. Fifteen minutes later I saw seven birds, six of them paired. As the display continued, more birds gathered until I saw a dozen or two at the same time, flying in close formation in pairs, threes, and fours. Around two hours later a long loose string of forty to fifty ravens approached, flying from the north at well over 1,000 feet elevation. They were in a straight line, not circling or diving. The displaying ravens drifted out of view to the southeast with the newcomers, but about twenty minutes later I saw birds approach from the north; fifty-five of them were visible at one time as they started soaring and doing flight acrobatics while gradually drifting off in the same direction as the others. Bernd Friz, the head forester of the Hahnheide, assured me that only two pairs of ravens nested in the area; he knew of no local roost, and he had never before seen a similar flight here, although he had once seen a flight of 150 ravens in early winter some twenty miles to the east.

I conclude that although some data indicate ravens act independently as individuals, other observations show that they have a strong attraction to each other and sometimes wander in groups. Nothing more can be said about the flock movements because so far they have never been followed, and they have not been seen regularly enough to detect any consistent pattern of movement. There is reason to believe, however, that some of the birds who fly together also roost together at night, and both traits can have large consequences for my questions of if, how, and why they share.

COMMUNAL ROOSTS

BALD EAGLES, quelia finches, ravens, red-winged black-birds, and fruit bats are quite diverse animals: But they and many others have at least one behavioral trait in common: They sometimes (most species only in non-breeding season) have communal nocturnal roosts. Individuals or groups fly from the roost each morning and return to it each night. A communal roost may have as few as ten birds, but it usually contains many more. The famous crow roost of Fort Cobb in Oklahoma is estimated to contain "at least eight million" birds at peak occupancy in January. Another, in central Kansas, is reported to have up to ten million. But Richard Weffesten, a Fish and Game employee, admitted: "I don't know how many there are for sure, but when you get into a roost that size a few zeros more or less don't have any meaning. It's beyond comprehension, no matter how you figure it." The evolutionary or adaptive meaning of the communal roosting is also beyond comprehension. It is still an issue of scientific controversy, especially as regards ravens.

I once saw a crow roost of a million or more birds at a woods on the outskirts of Munich. I had been looking forward to seeing the ravens that a biology student in northern Germany claimed were all over the Bavarian city. When I got there, I saw rooks (*Corvus frugilegus*) almost everywhere throughout the suburbs. About an hour before dusk an un-ending "highway" of these crows flew past near the main train station in the center of the city, right next to the Zoological Institute, where I was giving a lecture. They flew by toward the west in the same track, over and between the

159

high-rise buildings, and as it was turning to dusk an hour later, they were still coming by. I was told that they came to the city every winter from Poland and other Eastern Bloc countries. (A similar phenomenon occurs every winter in Berlin and Vienna.)

The crows were going to their night roosting place in a spruce grove just at the edge of the city, at Aubiger-Lohe. A friend took me there, where he had seen them gather for "at least thirty years." But when we arrived at dusk, not a bird was in sight! We were ready to leave when a thick dense cloud of them appeared (they had probably gathered elsewhere, at pre-roost staging areas). The cloud circled, banked, dove, rose again, and swirled. There must have been tens of thousands. From their calls I knew they were a mixture of three crow species: rooks, *C. frugilegus*, carrion crows, *C. corone*, and jackdaws, *C. monedula*. It was an awesome sight. Then another cloud of them came, and another. The vast clouds flew around independently, sometimes almost colliding but swerving apart at the last moment. Soon there were hundreds of thousands. It was getting dark, and the many birds made it darker. It was like being at the bottom of a tank, where ink was dumped in and swirled around in random directions. Eventually the horde started dropping down into the black spruces, and still more of them were coming in. It could have been a million, or two or three. I had no way of knowing, but there were no ravens.

Here in New England there are no crow roosts of such spectacular proportions, and I have never seen or heard of a blue jay communal roost. Nevertheless, the hundreds of thousands of "blackbirds" that gather near Burlington, Vermont, are impressive enough. Before I knew about this roost, I had on several occasions pulled over to the side of the highway to watch, awestruck, as thousands upon thousands of chattering birds flew overhead in the late fall evening—all heading north! These were a mixed crowd of bona fide migrants: redwinged blackbirds, cowbirds, grackles, and starlings, all either solitary or semisocial birds, combined into an endless, noisy, black, undulating chain that stretched to both horizons as far as the eye could see. Following their invisible skyway I tracked them to their roost, a huge cattail marsh. Here other similar avian highways converged, not only from the north but from many other directions as well. In the dawn the black throng rose like a giant black cloud, which again strung itself out into long chains. Ultimately, the noisy crowds landed in the distant forests and fields where they foraged like phalanxes of army ants. Later, near midwinter, when the snow becomes deep, the roosts dissolve, as the birds presumably congregate at staging areas farther south.

All crows, genus *Corvus*, probably congregate in roosts during some seasons or parts of their life cycle. Although most roosts of corvids are of single species, on occasion they may contain several. Raven researchers Rolf Hauri from Längenbühl, Switzerland, and W. A. Cadman from Wales cite ravens, carrion crows, *C. corone,* and jackdaws, *C. monedula,* roosting together in Switzerland and Wales, respectively. British Major-General H.P.W. Hutson observed a communal roost of ravens, kites, and starlings in Iraq. (No mixed corvid roosts have been reported in America.)

The general locations of some of the larger crow roosts have been known for centuries. By contrast, the roosts of several thousand to tens of thousands that I have seen in Chittenden County near Burlington, Vermont, are always temporary. The birds never stay in any one place for more than about a month. I suspect that there is a positive self-reinforcing feedback between the size of a roost and its permanence.

At night a great horned owl will sweep through a crow roost "like the Angel of Death," as John Madson has put it, and once an owl has located a particular roost, it might be dangerous for the birds to remain. Of course, if the roost contains hundreds of thousands or millions, each bird in it is still relatively safe. Even if an owl kills ten birds a night in a roost of eight million birds, each bird has a nightly chance of only one in eight hundred thousand of being killed. These are small odds indeed and probably better than moving to a less populated spot.

The bigger the roost, the smaller the predation risk to the individual. But there is a limit to how many mouths can be fed in the same area. Inexperienced birds may find better food sources if they follow experienced foragers from the roost, but this advantage has a limit; if the roost exhausts local food supplies, it is better to leave and forage individually or to join a smaller roost elsewhere.

Since there is much literature suggesting that communal roosts serve as "information centers" where unsuccessful foragers follow successful ones to food, I used the nearby crow roosts in Burlington to see if food discovered by one crow would quickly draw many. I made a surreptitious offering of a cut-up goat carcass on the municipal golf course within a half mile of the roost. Two or three crows came to feed, but that was all during the week (in December 1984) that I checked daily. A repeat of the experiment the next year was the same. Perhaps the birds didn't like goat? I repeated the experiment with piles of dried cracked corn. No difference. The crow roosts may somehow serve as information centers, but these quick experiments didn't give any indication of it.

Few birds exhibit such a variety of roosting behavior as ravens. Most of the published reports are anecdotal, so the meaning of the different behaviors is unclear, but some glimpses can be deduced from the different patterns that have been seen.

The most detailed and systematic information so far on communal roosting in the raven is the result of the inadvertent provision of a site by humans—a 500-kilowatt power-transmission line over long treeless expanses near the Snake River in Idaho and Oregon. By far the largest raven roosts ever recorded—over two thousand ravens—have been found on some of the towers holding up this transmission line. The birds leave considerable accumulations of feces on the insulators, which cause power short-outs costing hundreds of thousands of dollars each. This problem has sparked an intensive study of raven roosting under the direction of Leonard S. Young, a wildlife biologist, with the support of the Pacific Power and Light Company and the U.S. Department of the Interior.

Young and his co-workers color-sprayed, color-banded, and radio-tagged ravens and kept year-round records of numerous roosts on 380 miles of this line. They found, amazingly, that individuals moved from roost to roost, and that each of the seven roosts along the line had an apparent life of its own. The ravens, unlike other birds, convened in roosts all year long, but some roosts were typically winter roosts, fall roosts, or summer roosts. In general, roost occupancy increased greatly in August and September, when the current year's young joined, and declined to lowest occupancy in the spring. Within this pattern of seasonal predictability there was considerable individual variation. For example, each year the Marsing roost had a few hundred birds in April, was unoccupied during June and July, and held a thousand or more birds in October. In contrast, the Initial Point roost, some fifty miles away, had more than a thousand occupants in June and July and only a few hundred in October. The change in numbers at any one roost was gradual. At the Initial Point roost in 1985, numbers increased an average of about two hundred birds a month from March through July.

The raven's roosting behavior suggests a similar trait in whirligig beetles, which forage at night and gather into crowds of up to hundreds of thousands in the daytime. Daniel F. Vogt and I once marked hundreds of these beetles on Lake Itasca, Minnesota, and found that each night they dispersed to forage over the surface of the lake, but toward morning they sought out and followed other beetles until they ended up in a crowd again for the day, but usually not the same crowd they had left at dusk. Nevertheless, there were crowds gathered at the same place each day because some beetles remained near the site and served as a focus to attract others. In both the beetles and crows, and possibly ravens, one of the functions of coming together is to find safety from predators in numbers. That advantage can be derived without assuming, or making, social ties.

The significance of raven roost occupancy remains unknown. The data suggest that individuals, or possibly small groups, wander around, probably following the food supply, and join others at convenient local roosts. Presumably as food becomes available in an area, wandering ravens stay to exploit it and then join any roost they find. When the local food supply gets exhausted, individuals are forced to search elsewhere. I would guess that the unusually large roosts in Idaho are possible because there is a plentiful food supply there all year long: grasshoppers and rodents in the summer, the cultivated grain wastes on extensive fields in the fall and winter. Possibly shifts in roost occupancy reflect annual shifts in food supply that vary from place to place.

Few other records of the western raven, *C. corax sinuatus*, are available. J. E. Cushing, Jr., has reported on a roost of about two hundred birds that gathered for at least nine years in the fall and winter near Tomales Bay, California. Richard B. Stiehl, during his Ph.D. thesis work on ravens for Portland State University, found a roost of several hundred that had been used "for at least 10–15 years, and possibly over 30 years" near Malheur Lake, Oregon. This roost begins to form in mid-October, and a peak occupancy of 836 birds was observed on January 4, 1977. The birds were tracked traveling into and out of the roost as far as twenty-seven miles north and south.

Other raven roosts are considerably smaller. Stanley A. Temple reports on a roost of ten ravens in the rafters of an abandoned hangar at the deserted town of Umiat on the Arctic Slope of Alaska. This roost is occupied annually from November through mid-March. Various authors in Great Britain and mainland Europe report roosts of twenty-seven to seventy individuals.

Raven roosts in the eastern United States are also relatively small. Vincent J. Lucid and Richard Connor from the Virginia Polytechnic Institute found a roost of 106 ravens at Mountain Lake, Virginia, on January 6, 1973, which was claimed to be "the largest aggregation and the only communal raven roost known to be reported in the southern Appalachian Mountains." Vermont ornithologist Frank Oatman told me about seeing fifteen to eighteen ravens roosting near Greensboro, Vermont, on December 8, 1974. Four days later the roost had increased to fifty-five birds, and another two days later the numbers had nearly doubled. All of the hundred or so ravens then disappeared the following day and have not been seen there since.

The birds from any one roost may come from a vast area. Richard B. Stiehl tracked individual ravens from the roost he studied at Malheur Lake in Oregon as far as 300 miles to the northwest, so if we assume this distance as the radius of a circular foraging area, then the minimum area the birds from this one roost spread into or came from is close to 282,600 square miles. On any one day there is considerable movement. Since Stiehl tracked birds foraging up to 28 miles north and south of the roost, we can consider 56 miles the diameter of the roost's foraging range. It therefore includes 2,460 square miles. Wildlife biologist Skip Ambrose from Alaska told me that in Fairbanks radio-tagged ravens make a regular daily eighty-mile round-trip as they commute between their roost and a landfill near town, suggesting that an even larger potential foraging range is possible for ravens from a roost than Stiehl's data suggest.

Adult territory-holding birds remain on their site all year, at least if they can, and it does not seem likely that communal roosters are adults. However, Stiehl reports some locally breeding adults leaving their territories in Oregon, presumably to join communal roosts. Furthermore, many birds flying in and out of raven roosts are in apparent pairs. Are the roostmates juveniles or adults? There is no direct evidence.

I already knew that there were many reasons why the recruitment I had seen could not be explained solely by gatherings of local birds who happened to be within earshot of a bait. I also knew that if I could prove that ravens recruit from or with a roost, it would be a new discovery. That bird roosts serve as "information centers" from which birds follow those who know where food is located is a favorite theory. But inexperienced birds could also follow *any* bird that leaves, presuming it leaves because it knows where there is food. This, however, is a form of information *stealing*, the exact opposite of the active *recruitment* I wondered about. The literature suggested (but did not prove) that roostmates were juveniles, and my observations would indicate that the crowds of feeders came from roosts. If both these suppositions were correct, it would be another piece of evidence to support my growing suspicion that it is only the juveniles who recruit to overpower the territorial defending pairs.

DO THEY COME FROM A ROOST?

NOVEMBER 25, 1986. It is Thanksgiving weekend, and instead of bringing a turkey up to the cabin, I am hauling about 500 pounds of beef and a 120-pound calf. I must once again see what happens when baits are discovered, and by whom they are discovered: flocks or individuals.

A friable crust over about a foot of snow makes the walking difficult. But by ten o'clock all of the meat is carried up and spread out in a loose pile in the woods. I will now watch with a fresh eye.

Mike, the game warden, told me about still another poached moose and a road-killed deer. Checking on the moose, I see that many ravens have been there since the last snowfall four days ago; the moose skeleton is picked clean, and the ravens are gone. There are no mammal tracks. In contrast, the calf carcass I left out the last time only three or four miles away is still fully intact. The deer carcass is uneaten as well. It will be bait on my next trip.

By 11:20 A.M. my huge meat pile has been found. A single raven flies over, then backpedals with its wings as if slamming on the brakes. It circles once and makes a series of high-pitched short yips, circles a few more times, and disappears as a speck in the distance against Mount Bald. There are three more fly-bys by a single raven later.

NOVEMBER 26. A raven comes by at 7:09 A.M. and makes no sound at all. It comes by again at eleven, this time stopping to perch and make a series of about ten typical deep quorks. In the meantime, a pair of blue jays and a pair of crows have

166

also discovered the bait. In the afternoon it rains hard, and nothing flies. It sleets and pours all night.

NOVEMBER 27. A raven is already in the vicinity while it is still dark, at 6:30 A.M. I again hear the low, almost non-chalant deep quorks, but now there is also a new sound: some hollow, gonglike calls. At 11:15 there is again a single raven, making the deep long quorks. Today is Thanksgiving, and I take a break to eat with friends who live a few miles down the road.

When I return three hours later, at 2:21 P.M., I hear ravens as I'm coming up the trail. It sounds like an agitated cackling. A fight? Five ravens fly away as I approach. How-ever, the snow shows only *two* sets of raven tracks that come within about two feet of the meat. There has been no feeding. I had expected that they would return within a half hour, but no raven showed itself for the rest of the day. Surely they will be back tomorrow.

NOVEMBER 28. The next dawn. Nothing. About three quar-ters of an hour later, at 7:15 A.M., a single raven flies non-chalantly over the bait. Again I hear the hollow, gonglike calls. The bird does not go near the bait. It is surely one that has been here before.

At 7:30 a raven flies around and appears to be looking things over closely. It stays for thirty-two minutes, calling almost constantly. It is clearly excited and makes short, high-pitched quorks. After I think it has left, I venture to the outhouse, but just then it comes back, and I crawl on my belly through the snow to the backdoor of the cabin. The bird did not see me. It stays for ten minutes on a perch just above the bait, descends to the ground and hops within two feet of it, but does not touch it. Before it flies off into the dis-tance, it makes a series of six to eight knocking sounds with a "thunk" at the end.

A half hour later I hear high-pitched quorks in the dis-tance, perhaps a mile or so away, and then the hollow, gong-like calls. At 9:55 a raven flies by briefly; it is not excited, only making deep quorks and one series of knocking. Just flying by to check? Two crows and two blue jays are feeding.

No other raven comes by noon, when I take the afternoon off for a walk in the woods. There, a pair of ravens flies over me silently. They are zipping so low over the trees that their wingtips occasionally hit branches. They are separated by well over one hundred yards, and their heads are jerking constantly this way and that as they appear to be scanning the ground.

DECEMBER 8. I do not get to camp until late at night, and it is quiet except for the occasional sounds, like distant rifle shots, of the frost cracking the trees. The stars are still visible, but a mist is settling over them. It could mean a storm.

The carcasses and all of the meat I had left out last time are *still* intact! I see almost no evidence of feeding, but there are a few fresh raven tracks. One or two ravens must have been at the bait today.

DECEMBER 9. I am awakened by the first dawn at about 6:30. It is overcast and starting to snow. I have not slept well. Subzero temperatures get to my bones, especially those in the feet.

By 6:45 A.M. I have built a fire and drunk some hot coffee. I'm reviving, but soon I'm very much alive: At precisely 6:46 a ragged stream of ravens comes flying in from the northwest. There are twenty in all! They are silent, and some of them perch in the trees around the bait. Others circle nearby and descend into the forest. It is not until twenty-five minutes later that I hear the first yells. Strangely, as many times in the past there seems to be only one chief yeller. Within a minute the first raven comes down to feed. Five minutes later all twenty are down.

From here on for the rest of the day, things settle into a pattern. First, it snows hard all day. Visibility is very low— I can barely see the forest on the other side of the clearing— but still *more* birds keep coming. By midafternoon they have increased to forty and possibly more. Because of the low temperatures throughout the day ($-7°F$ to $0°F$) the muscle meat is rock-hard, and so far it is left alone. Up to now only the much softer fat, lung, and liver are picked at. I saw only three birds leave with meat in their bills for caching. Apparently, the birds do not cache when there is only hard-frozen meat available. They normally cache fresh meat they can easily tear off, and now they only cache fat (which never freezes solid). Caching fat may have nothing to do with better storability, as one publication on food-caching in ravens suggests.

Today the birds spend most of their time directly on the bait. But they flew up all at once nineteen times, afterward dispersing in all directions. When they come back, they come primarily (thirty-five times versus nine times) as singles or pairs. Each time, just before they settle back in to feed and while they are settling, I hear a lot of high-pitched yells. Again, only one or two seem to yell. The others are silent. Usually the caller gives at least forty yells before a new feeding bout begins. Afterward, things are pretty quiet. There are very few trills, and only twice did I hear the xylophone-like knocking with the thunk at the end. Once I heard the gong sounds, like slow single hits on a hollow log drum (about once per second for about ten times).

Most of the group takeoffs were unrelated to any cause that I could see. But one was interesting. There were two crows who never fed when the pack of ravens was feeding. At 9.06 A.M. one of the crows cawed loudly several times. All of the ravens left the bait, and a number of them flew up to the northwest where the crow had called, circling down into the woods. They have seen a coyote, I thought to myself. Sure enough, two minutes later it appeared at the bait, but it looked very nervous and immediately trotted away. (It probably smelled my urine that I sprinkle there liberally to keep them off, and they usually stay off for at least a week.) The ravens in the trees near the bait completely ignored it. The crows called twice more later, but these caws sounded *slightly* longer and not quite so high-pitched to my ears. The ravens kept right on feeding.

Today was also a nice opportunity to observe the three different corvids—blue jays, crows, and ravens—using the same bait. Each time the ravens left, the blue jays (one to four of them) flocked down to feed within ten to thirty seconds. And

about two to ten minutes after the jays went down, the two
crows came and the jays left. Eventually the ravens returned,
and when they did, the crows left. When only three or four
ravens start to feed, the crows already leave and perch in the
trees nearby, waiting for the ravens to go. Sometimes the
jays are still feeding when the crows are down, but generally
the three species avoid one another at the bait. Today, however, at no time did I see a bird of one species chase one of
another. The ravens rule the bait, but they let the others
down, maybe because they can displace them so easily.

The ravens are not mutually tolerant. I saw three agonistic square-offs in the trees where one raven aggressively confronted another, who squawked loudly in agitation. I also
saw ten very vigorous aerial chases. In one of these one bird
chased another fifteen times around the clearing! In most
cases the chased bird left the area. What do these chases
mean? Is it play? Are they mate-chases? Or are they chases
to get rid of rivals at the food? Or are they really the mated
territorial pairs trying to evict trespassing juveniles?

The birds gather not only to feed; despite their fighting,
they also socialize. Twice I saw one bird on the ground in
the field far from the bait joined by a second and a third until there were six or seven in a knot. There could have been
no attraction except one another. These birds never squab-
bled.

At 2:50 P.M. thirty-seven ravens suddenly leave the bait,
as if on command but without a sound, flying over the hill
in the same direction the twenty had come from this morning. There was no vocal signal. Five more come by in a minute or so and land in the trees. They do not go down to the
bait. After making a few gruntlike, gurgling sounds, they,
too, fly northwest.

One raven remains! It sings and goes down to investigate
the bait, but it does not feed. It perches over the bait, trills
some twenty times and makes gurgling sounds. Another bird
comes close, answering with a few quorks. At 3:04 the two
(the territorial pair?) finally fly off together toward the east.
I had not heard trilling all afternoon until now; the yells had
predominated. The trilling can therefore not be a signal to
begin feeding.

DECEMBER 10. This morning before dawn I am up in the tiptop of a spruce tree, straining my eyes and ears to observe the ravens coming to the bait. Will they all come from the same direction? Will they fly as a flock? How will they vocalize? At 7:00 A.M. I hear excited short, oft-repeated calls, some one and a half miles away, and within a minute a half dozen ravens are circling low over the baits and trilling excitedly.

But neither the cut-open deer nor the calf at the other side of the field draw them down. Are they afraid of the meat they were feeding from yesterday? (I have rearranged it slightly after shoveling off the snow and chipping off the ice.) Finally, at 11:30, four hours and twenty minutes later than yesterday, they begin to feed.

No more than nineteen birds come today, even though over forty left last night. So this is how the numbers at a bait are regulated: If it gets too crowded and fighting increases and/or there is better food elsewhere, some birds who have been recruited simply don't come back. Certainly yesterday the picking here was very hard and crowded.

There seems to be an inordinate amount of fighting going on today. There are many chases with apparent aggressive intent (one bird clearly tried to bite another it caught up with). No mere play, this. There are fights at the food. At one point I see two pairs come in to land in the trees almost simultaneously, and right after that there is a chase, and one bird looks as if it is being pecked.

It gets warm. Temperatures are near 32°F. The meat is no longer flinty hard. And today the birds are again caching continuously.

In the afternoon I check two old baits put out November 4. The calf, now buried under snow and ice, is still untouched, and the meat pile close to the old raven nest, still has only two birds feeding there, as before.

Walking in the woods near Mount Blue, looking for the telltale snipped twigs on the ground where they may have roosted, I hear a raven's excited long quorks. They are almost angry-sounding, but they register surprise and excitement at the same time. Something is there! I go on the run to the calls just in time to see a raven take off. Another one calls

nearby. There, in the dense brush, surrounded by fresh coyote tracks (but no raven's track), is a dead deer! It has been
glazed over by ice and covered with snow, and a coyote has
dug up some of the head and neck.

On my way home through the woods, I met two acquain-
tances, Danny Proctor and his father, coming back from rabbit hunting. "Seen any ravens lately?" "Yes," said Danny.
"In fact, two weeks ago, about forty ravens settled as it was
getting dark in a patch of pine trees near my house." A roost!
Just what I was looking for. "It was such an unusual sight,"
he went on. "I'd never seen anything like it. So I checked
back the next night, but they weren't there." The one roost I
had found two years ago was also used for less than a week.

DECEMBER 11. I get up at 5:30 in night darkness, quickly
heat up the wood stove to make my coffee and Cream of
Wheat, and by 6:20 A.M. I'm running through the woods to
get to the sentinel spruce I have chosen to climb. By 6:30 I've
scrambled to the top. And it is none too early. At 6:38 I hear
the short, rolling, quick quorks about a mile away, and very
quickly after that a loose, strung-out flock of ten ravens comes
whizzing by me, directly toward the meat pile. It is still almost dark. I can't make out the time on the dial of my wrist-
watch without a flashlight. The birds are silent as they come
close to the meat. They look beautiful—so sleek, so fluid—
from up close at treetop level.

Within sixteen minutes, other groups of six, four, seventeen, and two pairs have also flown by. Each large group is
preceded by calls that I hear from about a mile away. By
7:09 two more singles and a group of three have flown by.
All forty-six are coming from the same direction, and all are
going to the same place. There can be little doubt that they
came from a nocturnal roost. I could not feel "higher," because I've learned something: The recruitment cannot be explained by a vocal relay radiating in all directions to alert
others distributed in an ever-widening circle, as one hypothesis suggests. If that were the case, the birds should
come from *all* directions, and they wouldn't come before daylight, before there is any activity here.

From up here in the spruce tree on top of the hill, I can

see the white top of Mount Washington, the smoke from the Rumford paper mill, and endless woods. All the towns are in the valleys, so you don't see them at all. You just see the wooded hills. It looks as if you own the world. Everything seems close, and how much closer it would be if I could fly like a raven! But no wonder ravens fly at treetop level to search for food: *I* can't even see the white ground through all the thick branches below me.

Today I see no chases and very little fighting. The number of ravens at the bait at any one time varies from about twenty to thirty-seven. The crows, as usual, do nothing but sit high in trees close together while the ravens feed. I show myself to them, and they caw long before the ravens could possibly see me. Again, as with the coyote alarm, the ravens take off instantly. Later, the crows caw six more times for no apparent reason, and the ravens act as though they do not hear! Apparently the ravens can detect differences in the crows' calls that to me sound very much alike.

Are the ravens in pairs? So far I have tried to score whether, when they are coming in to the bait after having been dispersed, they come in singly or in groups. Ninety-seven times I counted what looked like single individuals, and seventeen times what looked like pairs flying together. Often fifteen or more would be sitting in the trees. I saw none that looked as if they were sitting in pairs, the way the two crows routinely sit. It is too soon to tell whether these numbers are significant. Maybe it will be different in March. Good to have the numbers now, in case they take on meaning later in the context of other observations.

DECEMBER 12. I am again up in my spruce, relaxing in the early morning darkness, when I begin to hear an excited chatter of ravens in the direction of Houghton Ledges. It is one continuous, tumultuous uproar of trills, deep quorks, knocking-gurgling sounds, soft nasal "unks," yells, and probably almost everything else in their vocal repertoire. The tumult comes out of the dark, and within another minute or so the big black birds are streaking by at tree level over the black silhouettes of the spruce tops. The group is led by seven in a tight cluster, followed within a hundred yards or so by a group of about forty. They are so dense that I cannot make a direct count. But the number is close to those that came yesterday at about the same time, and they come again from the same direction. I think it is one of the most thrilling moments I have ever experienced—the time, the atmosphere, the sounds and the sight of those birds, and the certainty of having another piece of the puzzle.

At the bait the birds are extraordinarily shy. Do these birds post sentinels? I have never seen any evidence for it. But they are very alert to the reactions of others, even crows. A crow is alone in a tree while a group of ravens is feeding. The crow caws, but none of the ravens is alarmed. A repeat experiment is in order: I peep out the back of the cabin so that the crow will see me, but the ravens will not. The crow takes off—silently. Within a second all the ravens fly up and leave in a flash. I am amazed at the way the ravens use the crow as a sentinel to *leave* but never (in my experience) to go down to the bait. Neither the crow's caws nor its leaving as such are a signal, but very subtle variations of either that I am unable to discern make all the difference to the ravens. Given this subtlety that they can decipher even in another species, how can I hope to discern the probably much more refined communication among themselves?

Today, when there were ten birds, there were eight agonistic interactions per four minutes, and fourteen when there were thirteen birds. This is about eight times more intense than yesterday at the same bird density, and the amount of food is only halved! I'll wager fewer will come tomorrow.

I get some new insights into calling today, too. The *variety* of calls made during a flock flight, to begin with. The trills are made not only during food discovery. I also heard trilling and the xylophonelike knocking by two birds obviously playing together in flight just before the ravens left the area. Are these courtship sounds?

The trilling is possibly a male assertive call. I heard it in my tame male: When he was in the presence of a female and I provided a large new animal carcass to them, he would sidle up to *her* and trill, as if taking credit for the food bonanza. I also sometimes heard it in the absence of a carcass when he was displaying to her. He also trilled to me when I brought exceptionally attractive food. The same sound was heard in the wild when males were sidling up to females, or approaching apparent rivals. Clearly, it signals a high level of excitement. The semantic implication seems to be: "Look, *fantastic!*"—either referring to oneself or to what one is providing or taking credit for.

I have worked on the ravens continuously now for five days, four of them long before light in the morning, and the work has continued just as intensely for two to three hours after dark as I go over the data. It is hard to do the many experiments and try to keep track of all these things, because I still don't know what is really important, what will vary and why. I feel that if I miss something, it may be *the* critical observation. Also, the longer you watch, the more each additional day has extra meaning, because the data are always relative to what went on before. The more background you have, the more pregnant each *new* observation becomes. But the intensity and the information overload is getting to me. I think it best I leave tomorrow, after seeing what comes to the baits in the morning.

DECEMBER 13. My alarm didn't go off, and I awake with a start. It is already 5:20 A.M. The main point of my day is to see what flies in this morning from where, so I vault out of bed into the cold, quickly put on my clothes, and run through the woods to my spruce tree. No coffee. I'm out of breath as I clamber and crash up into the top of the tree in the dark.

It must be near 0°F today, and it is breezy. The tree sways. The sky is still black to the west, and I see stars. To the east it is turning yellow, salmon, and red. I predict there will not be many birds today. They will probably start to visit the new bait I put out on the other side of the hill.

The birds are at least five minutes late. The first ones come at 6:48, and only eleven come in all. There is none of the excited tumult of yesterday. Instead, these eleven birds, who come in groups of four pairs and three singles, almost seem to be loitering. Some of the birds, rather than flying straight like arrows as they did yesterday, are playing in flight, diving, rolling. They call very little. I leave my tree, unthaw myself with coffee, and check on the new bait on the other side of the hill.

The new bait had eight birds feeding on it at 10 A.M. The old, now smaller bait within a half mile of it, which has been there for two weeks, still has only two birds. The two other new baits—the calf and the deer—remain unvisited even though they have been seen by many of the birds.

A pattern is emerging now. The birds hunt for food individually. At night they gather at the roost. And from the roost, groups go out at dawn to feed at the best food sources. If one large food bonanza is clearly the best, then most or all go there. As the amount of food at a site becomes less, aggressive encounters increase, and individuals start to feed at other, less crowded sites.

I include here some additional observations and thoughts for amplification. As shown in later chapters, many of the birds at baits are highly vagrant, moving from bait to bait as individuals rather than as crowds or groups. It is therefore likely that as the food supply deteriorates at any one site, the individuals who receive the least food (the subordinates?) would also be the ones most likely to move on first, searching for other food and other roosts.

While birds may seek roosts both to find safety in numbers and to be led to food, the existing roost occupants may gain by advertising roost location to recruit more members to dilute further individual risks like predation at night. Stiehl describes flights of many ravens soaring together near a roost and speculates that these advertise roost location, so that a roost can grow. Lucid and Connor and I have also seen soaring and flight acrobatics before ravens settled into a roost. And I have heard yelling and trilling at the roost in the evening that sound very similar to that at baits. It appears as if ravens have at least three ways of attracting recruits to a roost: yelling as at baits, encouraging others to follow, and broadcasting visual and aural signals from high altitudes.

All but one of the ten ravens roosts in Maine and other parts of New England that I have seen or that have been reported to me contained few birds (less than fifty), and they lasted only one to six nights. (The one exception is a roost in Strong, Maine, which is strategically located between three landfills at which the birds forage when not utilizing carcasses in the forests.) How can the apparently vast differences between the raven roosts (especially between those of the western and eastern ravens) be reconciled? I speculate that it relates to food supply.

A hint of how roosting may be tied into feeding can be seen from the following three sets of observations: (1) On several occasions I have observed a communal roost forming near a food bonanza (within a half mile) after crowd-feeding begins at that food bonanza; (2) the number of birds at the nocturnal roost is roughly similar to that at the nearby bait in the daytime; (3) as any one bait becomes depleted, fighting ensues, numbers of birds decline, and the nearby roost dissolves.

We can now tie all of these observations into a single simple model. First, in the East the roosts are highly erratic in location and ephemeral in duration because that is the nature of carcasses utilized by ravens. In the West, the birds presumably also follow the food supply, but the food depends on the local productivity of the land, rather than on the chance appearance of carcasses. Food is also much more abundant, and not easily depleted. Therefore, numbers of birds in any locality keep increasing or gradually declining, depending on whether the grasshoppers, rodents, or the corn crop is gradually shifting in abundance.

These relationships between distribution of food and roosts probably also explain the seemingly different social system reported by some European workers, principally in Switzerland. The Europeans worked at large permanent dumps, and most likely there were fewer carcasses available in the forests for alternative food. Hence there was less reason for the ravens at any one food site to leave. Although I provided carcasses or meat piles frequently at the same location, my studies mimic the natural situation; I usually removed the bait or allowed each bait to be *depleted*, that is, fighting occurred, and all but the most dominant birds eventually had to disperse.

TO CATCH AND MARK A RAVEN,
OR TWO, OR MORE

> *Ah, distinctly I remember it was in the bleak*
> *December;*
> *And each separate dying ember wrought its*
> *ghost upon the floor.*
> *Eagerly I wished the morrow;—vainly I had*
> *sought to borrow*
> *From my books surcease of sorrow—sorrow*
> *for the lost Lenore. . . .*
>
> —EDGAR ALLAN POE,
> *The Raven*

A GOOD FRIEND and colleague of mine is enthusiastic about my results with the ravens so far. But he has the temerity to say, "Bernd, you *have* to mark your birds so that you can identify them." In my studies on bumblebees and water beetles, I had marked individuals to discover what would otherwise not have been possible. But to catch *ravens?* I already felt extraordinarily lucky to have come close enough to *see* them. Wildlife biologist Lenny Young in Idaho had already warned me of the problems. He had tried to bait ravens to traps with every delectable morsel imaginable—cut-up deer, pronghorn antelope, cows, rabbits, pheasants, and potato chips. And he got no response for weeks. Black bear researcher Craig McLaughlin from the Maine Department of Fisheries and Wildlife had also told me that although ravens will clean up fifty pounds of bait meat in one day, they "do not touch bait if one puts down a black feather, or a dead black bird such as a cormorant with its head tucked under its wing." For me, they had also stayed away from baits without these props, but there is no doubt they feed; the only problem is, they are always cautious. Will they walk into a trap?

I spent several weeks mulling it over before deciding that, yes, it had to be done. The ravens must be caught. The only questions were how, when, and how many.

DECEMBER 23, 1986. Several possibilities have been considered—cannon nets, remote-activated spray marking, leg-hold traps. I eventually settle for trying a large walk-in trap to be camouflaged in the woods. I renovate the aviary where I used to confine Bubo, my owl, when visitors came to the hill. The trap is a 12-by-12-by-7-foot chicken-wire enclosure. One whole side serves as a trapdoor. I will prop the door up on a pole, and when twenty to thirty ravens are feeding in the cage, I will yank out the prop from my concealment in a blind, and the door will slam. I want to get many birds *at once* in order to obtain meaningful data. And the trap had better work the first time, because I cannot count on getting a second shot. I have tried the mechanism several times: The door slams shut nicely.

Now all I need is patience, and I keep my fingers crossed that the radio transmitters and colored wing markers I ordered will be here soon. Strange, a year ago I thought it would be utterly impossible to capture and mark these birds. Today I am confident that within a month I will yank the wire and capture at least twenty birds!

The snow is well over a foot deep now. It is truly the depth of winter, and the predators are probably hungry. A short-tailed weasel is luckier than the rest. It has found my 500-pound meat cache under the deep snow and tunnels all around it. As I come near, the small white face with coal-black eyes pops out of a hole in the snow. The weasel stands up on its hind legs with its forepaws hanging and looks at me. Then it bolts off. For fun I run after it, to find out how easy it is to run down a weasel in the snow. It is not very easy, and I return. The weasel seems not to have been greatly alarmed by the chase. Within a minute it comes bounding back. The animal is an immaculate white, even whiter than the snow, but as it comes obliviously right past me near my feet, I notice that the fur at its hind end has a lemon-yellow tinge. The tip of its tail is as jet black as its eyes. Back into one of the holes it darts, presumably to continue to feast on the 500 pounds of meat.

DECEMBER 25. I awake shortly after 3:00 A.M. Raindrops are pounding on the roof! Dawn hardly seems to come today. It stays dark, and the rain falls in sheets, quickly turning to ice. All of the trees are glazed, and by ten the birches are bent over double. In between squalls the breeze makes the ice on the trees tinkle and rustle. The birds are silent. Suddenly, overnight, all of their food is locked away under ice. For them it could be a disaster. For me it is merely inconvenient; I won't be able to go spend Christmas with my family. You can't drive on glare ice.

A raven comes by briefly at 11:10 A.M. It perches in the red maple near the cleanly picked-over cattle heads, now glazed with an inch of ice. It ruffles its feathers, calls a few times, and flies off. It ignores the three piles of meat by the trap and inside it.

DECEMBER 31. A crisp clear blue day. A male chickadee calls the first "dee-dahs" of spring, and a crow (presumably a female) makes a series of knocking sounds. The drive to Maine was uneventful; I did not see a single raven in all of the two hundred miles. Usually I see one, sometimes even two.

I am not sure what to focus on this time. Mostly I need to check out the trap, make sure it is still baited, and see if the ravens are feeding yet. Other than that, I will trust to chance. One can never plan too closely, because one never knows what one will find. Progress often depends more on how well one follows the situation than on how well one controls it. Especially when control is difficult.

About twenty ravens fly up when I get near the trap. They have eaten almost all of the meat outside it and even started to feed just inside the door. I think they are as good as caught already! Now it is essential to keep the trap continuously baited until everything is ready to spring it. If they leave, no ravens could be around again for months. But the raven roundup must remain for the future: I still do not have the radio transmitters or the materials and supplies necessary to mark the birds.

JANUARY 1, 1987. The party till past midnight last night with friends and students who had come to the cabin doesn't excuse my staying in bed this morning. I don't stumble out until 6:00 A.M. to begin my vigil in the top of a spruce. I expect that most of the birds will come from the same direction as they did the last time. In front of me, in the east, the dawn is showing. It looks like a giant fire at the horizon. First bright red, just above the black spruce tops, then orange above the red, then yellow, and yellow grading into light blue to dark blue to almost black behind me. The colors are pure, brilliant, and at the same time subtle and soft. I am in a beautiful place to see the panorama, but it is not in the flight path of the ravens today. Indeed, almost all of the ravens seem to come from *other* directions than they did the last time, although I cannot be certain which directions they flew from. I had expected them from the northwest, and my eyes had been fixed in that direction. Suddenly there were many birds at my back.

JANUARY 2. All day yesterday and all night it was so quiet that you could hear a twig drop onto the sharp, glistening crust in the forest. The sky was clear, and in the evening you could see the stars, but a thin veil was creeping in—the sure sign of a storm.

After a few hours of sleep I'm up again, this time early enough to build a fire, warm myself and have a cup of coffee, and still be atop a spruce before dawn. This tree is a better choice than the one I used yesterday. It commands a view in all directions, and this time I'm more alert.

No stars are visible, and the sky does not lighten much even at dawn. Puffs of breeze soon sway my lofty perch above the forest, and a few tiny snowflakes whisk by. By seven there are more flakes, and the light puffs have changed to intermittent gusts. Now I'm shivering so violently that the tree is not only swaying but vibrating. I'm near the top, where the diameter at my feet is about three inches, and I can almost touch the top of the perhaps 100-foot-tall tree. I am not worried about the tree being strong enough to hold my 160-pound body, but I am concerned about becoming stiff from the cold and too weak to hold with one hand and take notes with the other. But I'm determined to stay, because the ravens have started to arrive. Besides, I love the view—the mountains to all sides and the vast panorama over the black spruces with white birches among them, the black skeletons of the maples, and the unfolding sunrise. As Erasmus said, "The highest form of bliss is living with a certain degree of folly."

There are no colors this morning. It is a stark, majestic scene from up on my perch. The landscape seems lifeless—except for the ravens who now fly in like black arrows aimed at their target. And they, forty-one in all, are coming from *four* different directions. Each group comes from a different direction at a different time, spread out over an hour. Two pairs, who are later than all the rest, come in at a great altitude, perhaps over a thousand feet. Did they come from very far? Pairs are supposed to be in "territories." Obviously they have no strictly defended boundaries. Maybe they flew high to avoid detection. As they come close to the target, they set their wings and bolt straight down, turning and twisting this way and that, pulling up just before hitting the trees. They bank sharply and circle gently before landing at the bait.

I've learned something that would have taken months of laborious work with individually marked birds to find out. I know that the birds at any one bait are *not* all from one roost. They can be from at least four different roosts. The feeding groups at any one bait are not necessarily composed of one coherent flock that stays together. This is important information if one wants to work out the evolution of their sharing behavior.

The few snowflakes and the wind are the beginnings of a blizzard. Throughout the day the wind gets stronger, roaring and whistling through the trees, and the snow swirls and tumbles. The ravens again do little if any caching. The meat is frozen solid and the birds are content to tear or hack off just enough to eat. As before, they come and go in waves; there is either a group of them feeding or none at all. There are no more than fifteen birds at a time by noon, and after that they become fewer and fewer. One by one they fade into the wind, are lifted, almost thrown upward, and disappear in the distance. Many head toward Hills Pond, the direction from which a squadron came this morning.

It has been dark now for many hours. The wind is still howling, swishing, occasionally shaking the cabin (I am no longer "living" in Camp Kaflunk, having moved into the log cabin I built near it, which was christened Camp Believe It at the annual sheep-roast party last fall.) The snow tinkles lightly like sand as it is driven against the windows. My fire is warm, and I'm pleased that no smoke comes down the chimney, but the temperature is only 40°F six feet from the stove. Last night I made a couch with a tall back out of spare boards and put a mattress and blanket on it. It is in front of the stove, a good place to relax and listen to the storm.

JANUARY 7. Not one raven was seen all the way on the afternoon drive back to Maine to check on the trap. But there are plenty here. Most of the five hundred pounds of meat has been removed, and there are raven tracks (but no coyote tracks) patting down an acre of snow near the cage. The otherwise fluffy snow has been stamped down into a solid surface by thousands of overlapping bird tracks, which have stained it a dirty gray.

I have thought a long time about what "experiments" I will do in the next few days. Ultimately, I decide that the priority right now is to keep luring the birds into the trap. In the meantime, I can continue to count birds and their times and directions of arrival in the mornings, to finish documenting the idea that the crowds are birds coming from one or more nocturnal communal roosts.

JANUARY 8. I'm barely up in my spruce at 6:30 A.M. when I see black silhouettes shooting by against the light of the first dawn. I've never seen them this early before. Thirty-three

have arrived by 6:43, and by 6:54 there are fifty. Almost all
of them come from the southeast, a direction from which few
had come before. A new roost. They make almost no noise,
none of the raucous jubilation I heard the last time they were
coming to a *full* dinner table. There is little food here, and
by 7:00 most of them have already dispersed in all directions.
Only about fifteen stay to feed.

At 9:30 I replenish the almost depleted rations by the cage,
and ten minutes later two ravens feed on the soft, unfrozen
meat and repeatedly chase away others who are trying to get
at it. Then there are a lot of juvenile yells from a bird in the
forest nearby, although the two feeding birds are totally silent. Ravens start to gather from all around, and by 10:50
fifteen to twenty birds have descended to the bait.

It is now time to worry about replenishing the meat supply, and I make a run to the slaughterhouse in Readfield, returning with my jeep crammed full of about six hundred
pounds of meat scraps. The ravens are already done feeding
for the day, and fifteen of them soar and circle high over the
hill, at intervals diving down, tumbling and plummeting in
groups of two and three. Then up they go again, to repeat
the pattern over and over.

JANUARY 9. Since I hauled up a generous new pile of meat
yesterday afternoon in time for the ravens to see what they
might feed from tomorrow, I expect there will be many birds
coming this morning. What excitement then to see about
fifty of them indeed arrive at dawn in one large flock. I can
again hear them from at least a mile away. They are not
silent like they were last time, when the meat was almost
gone. As before when expecting fresh meat, they fly *noisily*
toward their goal. Do the ones who know about the meat betray their knowledge through their emotions, beginning to
call even before they leave the roost, so the others follow?
This would be the simplest hypothesis to explain active recruitment from the roost.

Only two come from the southwest, which is the direction
of Hills Pond, where the pair nests.

In the afternoon I rest, taking a short nap to the sound of
raven music in the background. There can be none finer. I'm

very tired from getting up so early every morning, and I must gather strength for the night's work of hauling up still more meat.

It is a clear, cold, wind-still night. The half-moon illuminates the snow like a lamp. Shadows are sharp. The snow crunches under my feet; it is so deep that it is a big effort just to walk. I not only have to walk, I must drag a toboggan up a 30–40° incline for half a mile with about one hundred pounds of meat on it for each load. After the third load, I'm drenched in sweat. All four layers of clothing are soaked. But after I'm done, when I can enjoy the luxury of dry clothes, drink a cold beer, and sit in front of the fire, I have the pleasure of knowing that the cage is well stocked for another week. I have come a small step closer to having color-marked, radio-tagged birds.

JANUARY 10. Getting up at 5:30 A.M. and climbing a spruce tree to watch the sun rise is getting to be addictive. There is magic in the morning. Last night I woke up about 2:30 to the sounds of a howling coyote. Now, in the northerly direction of Mount Blue, about a mile away, I hear a beautiful coyote concert. It sounds as if dozens of them are chiming in to howl, yip, and bark. While it is still quite dark, I also hear finches. They seem to call whenever they fly, and this morning they are flying in the dark. Their calling undoubtedly keeps the flock together, and they start to forage early to get enough fuel to last through the night.

At 7:00, five ravens finally come and perch in the trees. Then another raven comes, making very deep and insistent quorks, and all leave. Later on, these or other ravens only fly by. Eventually one discovers the new food I had put out during the night, and later I hear yelling. After that, birds start to accumulate at the bait. I can relax again. They are still here, and they *will* be caught.

Before leaving, I take a walk in the woods. The snow is deep and so is the silence, with the notable exception of the ravens. In all directions a mile or two away from the bait you can hear their "singing" with all the varied sounds of their repertoire except one. The juvenile yelling that is prominent at the bait is lacking. In an hour and a half I see at least

six pairs fly over toward the bait; quite a number also fly in
the other direction. Many observers have reported that ravens in crowds associate in pairs. This is strange, because
mated pairs are territorial. In one pair that flew over I saw
one bird flying slightly above the other, making sounds like
a mixture of running water and mellow knocking. The bird
just below was silent, but it seemed to be showing off with
flight acrobatics, diving and twisting and turning as it went
along. These birds look and sound as if they are playing and
socializing, and their playground extends for miles around
the bait.

I go next to check the calf carcass down by the lake, about
ten miles from here, where I had seen a pair of ravens earlier.
The calf was totally buried some weeks ago, but today the
meat is exposed—a coyote has dug it up. And predictably
there are ravens near it now. I see only two, however.

By eleven, it starts to snow very gently, and in another
hour it is coming down hard. As usual, I have waited until
the last minute to return home, and not even my four-wheel-
drive jeep can plow through, so I have to turn around and
come back. I feel happy, however, as I walk back up the path
to the ravens.

About twenty-five of them fly up from the bait into the
blizzard. I realize immediately that this might be a good opportunity to observe the remote possibility that they call in
others to help dig out the bait. If this is so, they should call
more now than they do when it is not snowing. Luckily, I
have a tape from one hour of recording yesterday. I can
count the number of calls and compare them with today. In
an hour I have counted only 520 vocalizations in the vicinity.
In contrast, yesterday in forty minutes, at the same bait but
with fewer birds, there were 1,625 vocalizations. Also, today
while it was still snowing hard, they all left by 2:45, although
they stayed until 3:30 yesterday. I think I can put to rest the
idea that they stay near the bait to keep snow from accumulating on it, and that they call in others to help them dig out
the food.

JANUARY 11. Yesterday's snow seems tame now. That was
only the lull before the storm! The blizzard came from the

north at night. It sounded like the roar of a giant surf, and it shook the cabin. The pots of water on the stove are frozen solid when I get up. Even my clothes next to the bed are under piles of snow, because the blizzard is coming in horizontally through the cracks in the walls. You know winter has finally arrived when it starts snowing inside a cabin that you thought was well chinked. Still, relative to the violence outside, it feels nice to be inside this little capsule of comfort. How can those tiny finches that I heard yesterday morning flying by in the dark survive out there? What happens to the even tinier kinglets, brown creepers, and chickadees? Where do the ravens roost on a night like this? How can they possibly hang on to the branches?

Even in Alaska, ravens in the winter rely not on impressive physiological or insulation adaptations, but on continuous heat production. In this cold wind they must have used up a lot of energy shivering to keep warm, and they will be hungry today. They will come early. But all the food is buried under deep snow. What will they do now? I could dig it up for them, but this morning I will leave it buried. It will be a good test to see if they can get at it themselves. So I climb a tree to watch as usual. But the thick horizontally driven snow makes it extremely difficult to see. I soon start to shiver violently, and I recall an article entitled "Hypothermia: A Quick Killer." To save your life it said to "clasp your arms tight against your sides and draw your knees up." Not good advice when you are in the top of a tree!

Well, the blizzard did not stop them, and neither did it stop me. I got some interesting observations. Rather than coming in one or several bunches, they come in one's and two's mainly, strung out over two hours. They fly very high over the trees and dive straight down to the bait. An impressive thirty-eight out of sixty-one that came in were in two's. Again, why are they *paired?* Only seventeen came singly. Although I presume most of these birds are juveniles, I wish I really *knew.* Nobody has that data. And I distrust speculations; I've made too many of those myself.

Only about a two-pound piece of meat that the wind keeps cleared is showing through the snow and by 7:40 A.M. the birds leave. No digging.

Due to the intense blizzard I am not in the tip-top of the nodding spruce today, but only half way. And at 8:45 I jump off my perch directly into the deep snow and crawl to the bait. (It is not possible to walk when you sink down *over* your hips with each step.) About twenty-five birds fly up. Thus, about thirty-five birds had left after inspecting the site. There is almost no evidence of digging. All the snow-covered meat in the cage as well as the two other meat piles in the field has been totally ignored. Only the small exposed piece of meat is being fed from. This does not mean they can't dig—I've seen them do it. Do they know the snow is deep? Maybe they will check out other sites first before returning here if they have to.

I love a blizzard. This is one of the best I have ever seen, and on top of that the ravens make it an unparalleled treat; the unusual situation has given me new perspectives on "group" flights to bait, bait retrieval, and regulation of numbers at baits.

JANUARY 15. This is becoming something like a military campaign, and I am like the general planning every detail. The outcome of the round-up may well determine the extent of my research for the next several years. If I am very "lucky"—if I have planned every detail precisely—I will take many raven prisoners on Monday morning. A small mistake that is unanticipated could ruin the whole thing. If, for example, I made the mistake of using a *rope* rather than a wire to yank the supporting strut from the trap door, the whole thing would surely fail. There would be a slight stretch in a rope, and that slight stretching would cause as much as a half-second delay between the instant that the birds are alerted and the time the door falls. Given their hair-trigger reflexes, that would be enough to allow them to escape. There are a hundred other details as well.

First, there is the issue of permits. I talked with Danny Bystrack of the Office of Migratory Bird Management in Maryland, to update my banding permit to include color tags and radio transmitters. The banding also had to be coordinated with the State of Maine at the Research Division of the Maine Department of Fish and Game headed by George

Matula in Bangor. The federal office told me about somebody else in Maine who had received permits to color-mark ravens, but checking with William Krohn in the Department of Wildlife Management at the University of Maine, I learned that the project was dropped. (The project leader was unable to catch any ravens.) If two researchers were using the same kind of markers on the same birds, we would have a problem. This also goes for radio frequencies. It would be disconcerting to think you were tracking a raven and learn later that the signal you were picking up came from somebody else working with other ravens or with black bears. So I cleared what frequencies were available, too.

Now it was a matter of getting the proper radio transmitters with the right frequencies, and borrowing receivers for those frequencies. Moira Ingle and John Persons in the Wildlife Department at the University of Vermont were just finishing their M.A. theses, for which they used radio-collared foxes and coyotes respectively, and I learned that their receivers might now be available. I was lucky: they were. I worked out a compatible frequency and phoned Telonics, the company in Mesa, Arizona. There I talked with Bill Burger, who was extremely helpful and agreed to rush the order; he would send it by express mail on January 13. (I got it on January 12.) But before ordering it, I had talked at length with Lenny Young of the Idaho Bureau of Land Management Office of the U.S. Department of the Interior, who is one of the two people in the world who has radio-tagged ravens. He told me what length and thickness and coating of antenna to get.

I now have the transmitter. How do you *attach* it to the bird? A backpack arrangement was suggsted, using two straps of tubular nylon sewed together with dental floss and sealed, to prevent untying, with superglue. Another shopping session. Lenny sent me the diagrams of how to harness up the bird with the radio. I studied it for any details I might have overlooked. Luckily, I had a dead raven in the freezer. I tried the harness on it. The strap was about two inches too short. Not a good thing to find out in the field. Another phone call, and another express package the next morning with more harness material.

Next the tags for wing markers. My bird-banding manual
listed half a dozen companies that make colored plastic from
which wing markers of various shapes could be cut. David
Capen from the University of Vermont's Wildlife Department, who had worked with ibises in Utah, suggested Herculite material. I called the Herculite Company in New York
and ended up in a long conversation with the sales manager,
Tim Pelton, who shipped me a number of sheets of white,
yellow, orange, and blue material, all free of charge, by express mail. But how to attach them? Again a long phone call
with Lenny Young in Idaho provided answers. Lenny told
me about the kinds of rivets to use, how to attach them on the
raven's wing and where. The color markers as such do not
provide information on *individual* birds until you paint numbers on the tags. What kind of paint will stick *permanently*
onto, specifically, Herculite plastic? More phone calls. "Vinyl
screen ink" will. It is made by Naz-der-KC in Chicago, Illinois. There is no answer at the Chicago phone number, and I
rush downtown, but I can't find any Vinyl screen ink in any
of the art stores. Another quick call to Lenny—yes, he will
send another air express package tomorrow and include some
paint for me.

In the meantime, I must locate a pop rivet tool, an awl, an
artist's paint brush, an opaque handkerchief, a roll of elec-
trician's tape, fine surgical scissors, thirty burlap bags, a
handnet with sacking for catching ravens (just in case it is
needed), several pairs of thick gloves, a ruler, a mosquito
hemostat, a flashlight, and a scale to weigh the birds.

This might be a one-time chance to get the sex of the birds
as well. It would be great to know the sex of the birds I am
dealing with. I recalled talking with Nat Wheelright, an
ornithologist at Bowdoin College, when I was there recently
to give a seminar, about sexing the birds by cutting them
open. I called him again to refresh my memory. It sounded
simple—you didn't even have to sew them back up—just
make a one centimeter slit, peer in at the gonads, determine
the sex, and let the bird go. I had sexed thousands of dead
birds before. Tiny ones. Ravens should be a cinch. But I'd
better give it a try, since with live birds I could make only a
tiny slit. I would practice on pigeons. A friend who has long

wanted to get rid of the pigeons infesting his barn came to mind. I bought a six-pack of beer, invited myself over for supper, and brought along my .22 rifle. With him holding the flashlight and me the rifle, we soon eliminated his pigeon problem.

The next morning a graduate student, Litia DiDomenico, and I tried our luck at sexing dead pigeons. You put the bird on its back and make the one-centimeter slit at the bottom of the last rib. *Sounds* simple. We *did* eventually determine the sex of one pigeon, but not before it was torn to pieces. Is there something special about pigeons that makes them so hard to sex? We could not seem to get the legs out of the way to get at, much less find, the last rib. There was too much flight muscle. The many feathers didn't help, either.

It will require a lot of good hands to pull this operation off. How do you get competent and eager help? You have a party. I put out a notice to the graduate students to come join the first annual giant "Raven Roundup" in Maine. We would meet at the Zoology office in Vermont on Saturday at noon, so as to get to Camp Believe It after dark. I stressed that it was important to arrive there only *after* dark so as not to disturb the ravens. We would all convene Saturday night at the cabin to get ready for the coming dawn when I would spring the trap. That was, *if* everything went according to plan.

Numerous things could already have gone wrong. Have the coyotes eaten all the bait in the last week? Has the deep snow made it easy for the coyotes to kill deer, so that all the ravens are elsewhere or not hungry enough to go into the trap? Will the ravens still be around? Will they go into the trap? Will . . . ? "If it works, it will be a first in biology," a colleague told me. But I had made good plans.

JANUARY 18. Just to be absolutely certain and leave nothing to chance. I called Charlie at Bowdoin and asked him to drive out to the hill to check on the bait and freshen it up with new carcasses. He went up Friday afternoon and called me to say it looked "like a massacre" had occurred there. Bits of bone and pieces of fur were strewn over an acre, and all the snow was patted down by thousands of raven tracks. No coyote tracks. Only ravens. And the birds had taken *all* of the five

hundred pounds of meat I had lugged up! Charlie left two
hundred pounds more. If he had not, I'm sure the ravens
would have been gone, feasting on a moose in Penobscot
County, maybe even up in Aroostook.

We were scheduled to come on Saturday. But I needed to
make *certain* that there would be bait—and birds. So I came
a day early. Alomst every scrap of the meat that Charlie and
his friend had brought up was gone when I arrived at night.
I put out another 150 pounds and waited in a tree at dawn to
see if any birds would return.

None came before 7:00. But the bait worked wonders.
Eventually thirty ravens arrived from the northwest, and
then nineteen more from the southeast. I was later unable to
keep track of new arrivals, but at least seventy were in the
area by three in the afternoon. It started to snow. They left
early.

Baiting the birds continuously to trap them is also part of
an experiment. I want to know how many birds will assemb-
le at a nearly infinite resource, such as this one has been so
far. The results already show that the birds feeding here are
not an exclusive club. Others keep joining and joining and
joining until, most likely, they are excluded by competition
at diminishing resources. Clearly this is not a kin aggregation. And now even here there is keen competition for the
diminishing resources—resources that are beginning to tax
my capabilities to supply. An "infinite" meat pile, with these
voracious ravens, is really beyond me. The feedback between
numbers and food can already be clearly seen—the excess
birds fly by and inspect but don't bother to drop in and try to
feed.

But the main reason for the prolonged feeding has, of
course, been to try to get the birds used to going into the enclosure. Like humans, they show no hesitation whatsoever
for the most foolhardy behavior, provided others are doing it,
too. If a group goes in the trap, the others follow as if that
were the only place to be.

I have already hidden myself in the blind where I will
pull the wire, and they still go into the cage. The blind is
"safe." Last night after dark I yanked on the wire to trip the
door, just to make absolutely sure it works. It does. All the

materials have been assembled. All the steps have been practiced.

Evening. It is snowing hard now. I'm depressed. Tonight I expected two carloads of graduate students from Vermont and one from Brown University. It was to be the "Raven Roundup" party, to get ready for the main event at tomorrow's dawn. There is no way that I could possibly begin to handle and process by myself the twenty or so ravens I plan to catch tomorrow.

This can't be happening now! Nobody showed up. Why? They will say that they couldn't make it because of the snowstorm. But as far as I'm concerned, there is no such thing as an excuse. They could have driven at twenty miles per hour. If you use excuses *not* to do things, you can *always* find something to blame it on. I also have no excuse. I overlooked something—a contingency plan for a snowstorm.

JANUARY 19. I sat for hours in gloom in the cabin during last night's snowstorm. But the dark veil suddenly lifted: I heard human voices! I rushed eagerly outside to see five flashlights blinking and weaving along the path through the trees and the still thickly falling snow. Moments later the loyal and brave crew all arrived at once, carrying large packs filled with bedding, beer, and bread. After shaking snow out of beards and packs, everyone gathered around the fire. What a relief to be at this place, in the woods, in the snowstorm. And with the upcoming excitement, we were soon in high spirits.

I had already laid out on the table all the tools and materials for the morning's work, and we had a review session to figure out what precisely each individual would do.

I would get up before daylight, build the fire, heat up the coffee, and go out and wait in the blind. The rest were instructed to lie low and under *no* circumstances peek out the windows (which were covered with tar paper), because the ravens were not to see anyone, anywhere. The crew would spring into action as soon as they heard the slam of the trapdoor, after I pulled out the strut.

"Twenty ravens, huh?" asked one of the graduate students.

He smiled. He also looked skeptical. "Yup. At least twenty," I replied.

The storm stopped just before midnight, and I went out once more into the trap to brush the snow off the meat. There was not much left, and it was all frozen solid. I debated whether or not to bring a bag of fresh bait from the cabin, but I decided against it because it seemed to me that the number of birds coming at dawn would *not* be a function of what food was out there, but what they *expected* to find; the number coming tomorrow at dawn was probably already set. The question now was, how many would gather inside the trap? I begin to figure: The meat inside the trap is solid, and it will take a bird some time to tear enough off to eat before it is satiated and leaves the cage. If I now put in chunks of soft meat, they could tear it off more quickly, so there would be much more traffic in and out of the cage, but few ravens in it at any one time. How many will go in with less than ten pounds of meat available? Fewer presumably than if I put in a big bonanza. I decide that the best strategy is to leave only the frozen meat.

The sky cleared, and the moonlight kept me awake. When the alarm finally rings at 5:30 A.M., I am up in a leap. One of the graduate students, Wolfe Wagman, is up, too, feeling the excitement. He makes real coffee as opposed to my usual instant solution, and we chat briefly by the fire. Then I run out, plow my way through the snow in the dark, and settle into the snow-covered blind. I test the tautness of the wire and kneel in the snow, straining to see the cage through my peep hole.

It is still, but I don't have long to wait. They come early. By 6:30 I hear the soft nasal grunts of approaching ravens. Then there is the beautiful ripping sound of air rushing through wings as they dive and bank down into the trees. Then the loud yells. The back of my blind is covered with a green tarpaulin, so that no bird can see any points of light *through* the blind, which would give away my presence if I move. But I do not move. With the firs and spruces densely woven upright all around and over me and covered with fresh snow, no one, not even a raven, would suspect this is a blind, even if they walked directly past it.

As I peer through peepholes, I can see dark shapes flying about in the gray dawn. They are coming closer through the trees, and they are walking on the snow. There is more yelling. Closer this time. A bird perches above me, looks directly at the bait and yells as if begging it to jump into its mouth. But the bait stays where it is and the bird eventually flies down. I can now see it and other ravens actually walking into the trap. My heart starts to pound. When to yank that wire? If I yank now, I'll get at least ten. To hold even one would be an unforgettable experience. But I'll need at least twenty to get meaningful results.

Ravens are now fighting in the cage over the picked-over scraps there. Will many leave quickly? I maneuver stiffly, to be ready to give that yank on the wire. A raven in a nearby tree makes very rapid strung-together, excited calls. In an instant the birds in the cage, and all the others around it and in the trees, erupt like panicked chickens in a coop. The sounds of heavy wingbeats are all around, and in another second every single raven is gone.

But all is not lost. Within two minutes I again hear the yellers, and ravens again come flying in from all sides. Through the tight lattice of twigs and snow I again see them going into the cage. They are piling in fast, and none so far is flying the other way. First there is a constant rush. Then it becomes a trickle.

The moment has come, but I'm almost afraid to yank. Will they all fly off again when I reach up and fumble with the wire? What if the trap *does* work, catching more than twenty large, hawk-sized birds with eight- to nine-centimeter-thick bills? Pandemonium. I expect nothing less, and I reach up and yank. The door slams, and I hear a muffled roar from inside the cabin. (The crew up there had been watching from peep holes.) But there is not a single raven sound.

I peer out my hole. Ravens are flying wildly about inside the trap—there must be close to forty. This is probably the most exciting moment of my scientific career, because now I have the means to get the data for solving this amazing puzzle that has cost me more time and effort than any other research so far.

The crew pours out of the cabin, and within a minute we

swing into action. As I enter the cage, or my assistant, Steve Smith, does, the birds fly into the far corner and pile up into a black, seething mass. All are docile, except that they bite when you grab them. We catch one after another by hand, put each into a burlap bag, hand the bags out, and carry all forty-three filled bags into a darkened room in the cabin.

We work nearly all day to process the forty-three ravens in assembly-line fashion: First we get the weight, then measure the bill and wing lengths, note mouth color, attach the wing tag and the U.S. Fish and Wildlife aluminum leg band, and finally we release each of them out the door. (They are all "left-wing" ravens. This is not a joke. It has practical significance for coding in case we do the experiments again next year, when we could use the same color combinations, but on the right wing.)

The processed ravens fly off on rapid wingbeats, squirt liquid feces, jerk their heads, and look back in all directions and remain silent. The attachment of the two radio backpacks—getting the straps just right and sewing them in place with dental floss—is trickier. We hope we've done it right. Two radios is all I could afford, and we put one onto an adult and one onto a juvenile. However, the radio-tagged birds will not be the main part of the study. They're merely a probe. The important anticipated data will come from the color-marked birds. (I often try to insert a probe for potential future research into the main experiment.)

We have been working for many hours, and few of us think about stopping to eat, except Jim Marden. Jim had brought along pork chops, and at about 1:00 P.M. he finally does us all a favor by frying some up. Meanwhile, at least three of the hopping burlap bags have gone flat as the occupying ravens escape. These ravens now find themselves in a strange world of talking, laughing, working people, and they are making a nuisance of themselves. At first we try to catch them; we fail, and they eventually quiet down. Then we leave them alone. One sits nonchalantly on a log near the stove where Jim is frying the pork chops. Two escape into the wood-storage room.

We enjoy pork chops on the run, and Jim offers one of the escaped birds in the wood room a pork chop bone. It is only a joke, but to his great surprise the raven grabs the bone and starts to pick at it. Even more surprising, its companion tries to steal it. A squabble ensues right in front of our eyes. This unbelievable sight convinces me that I could capture and house *wild* birds and study them in captivity. Maybe I could some day build a huge cage. . . .

The birds, after being released at the door, are no longer anonymous ravens. They fly out over the snow as Red #8 (R 8), or Blue #2 (B 2), or White #6 (W 6). Most (thirty) are red, the coding for birds that were born last March. The blues (seven) are the young from the year before that. And the rest of the birds, the whites (only six) are the adults, who are at least three years old and potentially able to breed. Here, for the first time, is direct evidence that crowds of ravens, *Corvus corax*, consist mostly of nonbreeders. It should be a solid and an essential piece of the puzzle.

As we are fastening the pop rivets onto the patagial wing disks, we use a little superglue for added reinforcement. One student from Sweden asks me what glue I had used for putting tags on bumblebees in my studies at this same place some ten years ago. It turned out that both of us had used the same resin, and it was not very effective. Now he uses superglue, and he says in his Swedish accent, "*This* stuff is goood!" and we both laugh. It is easy to get international agreement in science. Scientists all have the same standards— they are set not by beliefs, but by what works best. Of necessity, there is therefore universal unity. And unity makes for good will.

Note: One raven, Red #11, U.S. Band No. 706-21322, who was banded on this first raven roundup on January 19, 1987, was recovered in early April 1988 in Edmundston, New Brunswick, Canada, a straight-line distance of 220 miles north of the study site. (The bird had been bitten by a wolf in a large outdoor enclosure.) Red #11 is now in a cage at a zoo in St. Jacques, Canada. Another, Red #14, U.S. Band No. 706-21326, was caught in November 1988 in a trapper's set in Eustis, Maine, only 40 air miles north. One adult, W 4, continues to be seen almost every day two years later at the study site, and four other birds, R 7, R 21, R 26, and B 2, were each seen only one or two times two years later.

COURTING AND DISPLAYS

FINDING AND WINNING a proper mate is an important event, which many animals execute with considerable energy and inventive expertise. In many birds, especially those of the tropics, females assess a male's eligibility and usefulness for future offspring by some "useless" characteristic: having the ability, for example, to endure doing foolish and lengthy exercises without stopping to feed; making a lot of noise; possessing brightly colored but functionally useless feathers; or spending time courting while others must forage for a living. These characteristics are, however, an indirect measure of genetic quality, because they may represent overall vigor, which offspring would inherit. A female can, in choosing the male (or vice versa), also utilize his behavior to directly benefit herself or her offspring, rather than gamble that there is substance or some beneficial future promise behind his present extravagance.

Some females can raise a family without a male's help, but a female raven is utterly dependent on her male for food for over a month of every year. As one might expect in birds who rely on their mates to feed them and their young, evolution has provided safeguards in the courting behavior that help a female assess whether or not a suitor is going to be a good provider. It is done by a very "clever" ploy. During courtship the female acts like a helpless fledgling, mimicking both the behavior and voice of the begging young. The evolutionary rationale is that if the male can provide for her when she mimics a fledgling, he can probably provide later for her and hers as well. But this is not just a test, because the females in many diverse species need to be fed, and at *this* point in the evolutionary game neither parent can produce young *unless* the male feeds his mate.

Whether or not a male can actually provide directly or indirectly for his offspring depends on many tightly interrelated things, such as his vigor, the quality of his territory, his dominance, and in ravens possibly also his recruitment behavior! Almost nothing is known about the basis on which ravens choose mates. We can only make intelligent guesses based on the well-developed theory of mate choice as it applies to other birds. One thing is sure, part of the choice is, at least in the immediate sense, related to displays that draw attention to suitors.

In many birds, males and females are easily distinguished from each other by plumage, voice, and behavior. In ravens the sexes look alike (males tend to be larger than females, but there is overlap) and to our undiscerning ears and eyes may sound and behave nearly identically. Nevertheless, there must be differences, and I will try to unravel some of the confusing and not very well-known details of courting as it relates to ravens.

The sexual display ceremonies of the raven are described and illustrated in Franklin Coombs's and Derek Goodwin's books on corvids, but their information is derived primarily from Konrad Lorenz's observations of one pair in 1932. The German edition of Lorenz's book is illustrated with photographs, the English with drawings made from these photographs.

The other study on ravens' sexual display ceremonies that forms the primary basis for all textbook descriptions is *"Untersuchungen über das Ausdrucks- und Sozialverhalten des Kolkraben* (Corvus corax) *in Gefangenschaft,"* ("Inquiries on the expression and social behavior of ravens in captivity") the classic work by Eberhard Gwinner published in 1964. Gwinner worked with eighteen ravens, which he kept in several separate groups, or societies. Each group established a rank order headed by a dominant male, who was chosen over the subordinates by females. The birds bred and raised young in captivity, and from these studies we get an unusually detailed look at ravens' social behavior and home life.

The well-known scientists who have studied ravens—Oskar and Magdalena Heinroth, Gustav Kramer, Lorenz, Gothe, and Gwinner—all maintain that the male raven's dominance is established, reinforced, and maintained by his self-assertive display (*Imponierverhalten*). A male in the presumed first phase of this display stands tall with extended neck and bill pointed high. He depresses most of his head feathers but erects tufts of feathers (the "ears") just behind and above the eyes, slightly droops and spreads his wings to the sides, fluffs out his shiny lanceolate throat feathers and accentuates them by swallowing motions, flares out his flank feathers so as to appear to wear baggy pants, and flashes the white nictitating membranes over his dark-brown eyes. He struts stiffly about in a slow, deliberate manner. The female has similar self-assertive displays, except that her baggy pants and throat feathers are less pronounced. This self-assertive display is supposed to escalate to the *Dickkopf*, or "thickhead," display whereby the male raven fluffs out all the feathers of his head so that the ears are no longer seen.

Males give their self-assertive display at all times of the year, and Lorenz and Gwinner assign the same significance to it: It shows and maintains dominance by challenging rivals, suppressing a similar display in others, and impressing members of the opposite sex. It suggests macho power and daring to other birds and to most human observers. For example, Gwinner's Davida (a female imprinted on humans) courted him whenever he dragged large heavy objects past her in the cage, or when he worked with a hammer or ax, showing movements that evoked power, force, and perhaps daring, typical of a male courting raven. (Presumably in the female a similar display is meant only for rivals, not for the opposite sex.)

But male ravens also have an entirely different behavior. They make deep *bowing* motions to the female, during which the head feathers continue to be fluffed, the ears are no longer visible, and the bill is pointed down rather than up. Both Lorenz and Gwinner consider this an escalation of the self-assertive display. Can this be so? Another interpretation seems more likely to me.

In the *Imponierverhalten* the birds of both sexes are generally in exaggerated tall, *erect* postures. I cannot reconcile this with "escalation" to bowing ceremonies, where the birds *lower* their heads. Both sexes court with their head feathers fluffed out to the maximum degree, bending forward and down while simultaneously spreading their wings sideways. Putting their bills earthward, the males, apparently with great effort, emit guttural "tjo-gagh" calls; the female makes "rru-rra" or knocking calls. The Heinroths saw their captive male perform the bowing ceremonies with fully fluffed-out head feathers near a strange female, but during attack he stood *erect*, lowered his head feathers, and accentuated his raised "ear" feathers. Gothe's observations agree with mine and those of other investigators, that *threatening* behaviors are characterized by erect posture, "ears," closely compressed feathers, raised bill, and fixed gaze. Given all of the evidence so far, it therefore seems far more likely to me that instead of an escalation, the changed postures from erect ear to thickhead and then to bowing displays, represent a *change of motivation* from aggression to sexual interest. Perhaps a show of power is needed by a male for the *opportunity* to court. But by itself it may not win hearts. If this is correct, then mated *pairs* in the presence of intruders they wish to exclude should never "escalate" to the thickhead display. And this is what I observed in the field at the baits; pairs showed their ears only to strangers at the bait, and fluffed heads only to each other. The already mated pairs were showing off their dominance to the vagrants because they wanted them to leave, not because they were sexually attracted to them. This was confirmed when they were feeding by themselves; then their feather posture was neutral.

At my baits (see next chapter), fluffing out of the head in the absence of sexual display was clearly a submissive gesture. In the *Dickkopf* self-assertive display described by both Lorenz and Gwinner, the head feathers are also fluffed out, but body posture is entirely different—the male stands tall and erect as in the typical self-assertive display, rather than having his head meekly tucked in. Perhaps there are two signals in the fluffed head, *Dickkopf* display that say, "I'm impressed and unaggressive to you, but I'm dominant over the others, so I'm a worthy partner."

Lorenz and Gwinner presumed still a third "escalation" of the male self-assertive display where the *Dickkopf* male assumes a nearly horizontal posture and makes extreme bowing or retching movements, calling while crouching, spreading its tail and drawing its nictitating membranes over its eyes. Although this is indeed probably an escalation of the *Dickkopf* display, it seems to me one in which maximum *prostration* toward a suitor, rather than maximum domination, is being expressed. (I saw the same or a very similar display in females courting males.) The dominance expressed to the crowd has vanished, and the submission to a certain one *in* the crowd is stressed.

I saw at the baits that subordinate birds almost crouched, pulling in their fluffed-out heads in a more or less frozen display while in the presence of superiors. But other specific displays are used during direct confrontations, and they have different meanings in different contexts. For example, the infantile begging, during which the bird is in a partial crouch and calls and rapidly vibrates its wings (and sometimes also tail) like a baby bird, is used by the female when she is begging for food before and during the twenty-one days of incubation and up to two weeks after incubation when she is fed by the male. A similar begging behavior is also used as a submissive display in a confrontation. This display develops into a low crouch without bill opening and begging sounds that simulate the female's copulatory solicitation, but both sexes use the same display to acknowledge a superior's power. Again, bowing or crouching represents submission or respect, but not dominance.

In most animals, courting is seen (by humans) as an immediate prelude to mating. This may be true for robins or warblers, which do not have much time to waste after meeting each other on their breeding ground at the beginning of a short summer. But I suspect the situation is different for ravens, which do not migrate and which may continually associate with others of their kind for many years before they mate. In them, courting *per se* could be the culmination of a long process, as it sometimes is in humans. The beau that the teenage belle prefers is not necessarily the one with the deepest bow and the biggest smile. Perhaps his sex appeal also depends on how well he plays basketball or dances, or whether he wears designer jeans and has influential friends.

Do the birds size each other up before even beginning to initiate any overt sexual displays? There are several suggestive possibilities. One, the raven's spectacular acrobatic flight maneuvers. The birds typically choose updrafts where they ascend to a thousand or two thousand feet and then execute rolls and other maneuvers during headlong dives. Dirk Van Vuren of the University of Kansas, who made systematic observations of this behavior on Santa Cruz Island near Santa Barbara, California, found that 95 percent of the rolls were half-rolls. During a half-roll, the raven folds one wing back at the wrist, rolls rapidly onto its back, bends the other wrist and reverses direction to extend both wings. Half-rolls were performed to either side, and sometimes birds rolled back and forth in both directions. Three percent of rolls were full-rolls performed in a slow, steady motion with wings extended. One percent were double-rolls, in which two full-rolls followed each other in one continuous maneuver. Van Vuren twice saw "an Immelmann turn in reverse": the raven did a one-half inside loop on its back which it concluded by gliding upright in the opposite direction. Sixty-two percent of the rolls followed one another in sequences of 2 and up to 11. Rolls were interrupted only by brief glides of one to three seconds. One raven virtuoso performed "a sequence that included 6 half-rolls, 2 full-rolls, and 2 double-rolls." Each set of rolls was usually preceded by vocalizations. In some of these flights performed in pairs, the partners momentarily grasp each other's feet; they may also pass objects back and forth.

Many of the aerial displays are performed at great height, but Zirrer describes observations near his cabin in the Wisconsin woods of ravens assuming the shape of an arrowhead with wings partly open and diving like reckless daredevils straight to the ground. They halted just above the ground by spreading their wings and pulling back up to repeat the maneuver again and again to the accompaniment of "many croaking, clucking and gurgling notes."

Gothe describes four different kinds of flight that he assumes are mating displays. His *auf-den-Rücken-werfen* ("throwing

onto the back") corresponds to the rolls as described by Emeis
and Van Vuren. However, Gothe states that these flights occurred during most of the year, but seldom in May and not
at all in July and August, whereas Van Vuren saw them
year-round. The accompanying vocalizations included high
"gruh" calls and ringing "klong" or "djong" calls. Gothe
suggests that the rolls are not necessarily associated with the
Sturzfliegen (dives), which are performed at great height
by the members of a mated pair taking turns, unlike two
doing it together as an apparent pair as I have routinely observed in juvenile crowds. Gothe also describes *Schleifen-
fliegen* ("loop flight"), where mated pairs soar in circles or
loops high above their territory in January to May, with one
making "rrock" calls in rapid succession, to be answered by
the partner (presumably the female) in a high-pitched voice.
In *Gleit- und Wellenfliegen* ("gliding and wave flight") the
pair flies low over the treetops, the male close over the female
or ahead of her, making "rru-rra" calls and erecting his head
feathers.

What is the function of these displays? Van Vuren noted
that the frequency of rolling was the same in the fall, winter,
and spring and therefore rejected the hypothesis that the display was related to courtship, because in ravens breeding is
strictly seasonal. The display seemed to have nothing to do
with the number of other ravens (from one to five) with
whom the rolling birds were associated, so he also rejected
the hypothesis that rolling served as a "social display" and
decided that "play" was its most likely cause. However, although breeding is seasonal, courting takes place year-round,
and displays seem often to involve juvenile pairs. Thus, the
displays may function not only to reinforce existing pair
bonds, but also to assess mate quality in single birds.

Regardless of the time of year, many of the display flights
by pairs of birds take place when there are only a few other
ravens around as well as when there are crowds of forty or
more. Near one roost in Maine on December 19, 1984, I observed that in a soaring and aerial display by eleven birds,
there were five groups of two who performed closely side by
side in unison. On March 21, 1985 (in Vermont), I again

saw four "pairs" in a flight that also included five singles. Clearly these were not mated pairs, because the females are incubating in March. During the flight maneuvers, the two birds of apparent pairs were often so close together that it was nearly impossible to distinguish the individuals. Since almost all observers who have described the aerial displays of raven flocks talk about seeing "pairs," and since we also now know that raven crowds are primarily *nonbreeding juveniles*, it is almost certain that these are not adult pairs in the conventional sense; they are the raven equivalent of teenagers.

Ravens have other apparently "useless" behaviors besides flight acrobatics. Gwinner describes numerous variations of play in his captive ravens, including sliding on a smooth surface, carrying objects in the feet, and hanging by the feet and bill. The sometimes complex motor patterns initiated by one bird were often copied by its cage mates, and the performers were then attacked by higher-ranking ravens in the cage. Similarly, Richard Elliot saw a raven in the field with its throat hackles ruffled, like those of a sexually displaying bird. This apparent male was performing in front of another bird hanging by one foot, two feet, or its bill. A third bird attempted to mimic it. Elliot suggests that the behavior is a courting display by males. If it is indeed a form of showing off, it may explain why Gwinner's hanging ravens were regularly attacked by cagemates; other suitors would gain if they could eliminate the competition or do the trick better. Interestingly, hanging behavior is a well-known central feature of the courtship display that has been greatly elaborated upon by corvids' closest relatives, the birds of paradise. The intense competition between these males for females has produced spectacular visual, aural, and acrobatic hanging displays that are one of the great wonders of the bird world.

A raven's antics in its attempts to get attention as a prelude to courting can sometimes be amusing, especially when exercised on the wrong love object, as is common when the animals are raised without contact with their own kind. At least this is how I diagnosed one set of curious behaviors of Edgar, a captive raven kept by Catherine A. Hurlbutt, of Denver, Colorado. Ms. Hurlbutt wrote me that after some considerable effort (for seventeen years, since 1972) she managed to get Edgar to clearly enunciate "Nevermore." But all her work seemed to go for naught when, in 1988, a macho Viet Nam veteran came to board in the house. Edgar followed (and still follows) the man around like a puppy, completely ignoring his former keeper. He not only follows, now for the first time ever he displays in front of him a variety of "tricks": rolling on "his" back, often while simultaneously grasping objects in "his" feet, rolling spoons, clothespins and other objects up in paper and finally even crouching before him and vibrating "his" tail (the female mating incitation display). Ms. Hurlbutt feared she had lost her touch with the bird. But since Gwinner's captive female ravens were attracted not only to macho male ravens but also to male humans acting in an exaggerated and deliberate manner, I suggest that Edgar is really Edgarina. In the wild, of course, a raven has many more options to impress suitors.

Since ravens mate for life and may live a long time (ravens in the Tower of London usually live twenty to twenty-five years, although one, Jim Crow, lived to the ripe old age of forty-four), both sexes are likely to be highly selective in choosing a mate. There must be some way in which the birds evaluate each other on a relevant basis. In unmated ravens, courtship is a year-round activity, and the young birds begin it in the late summer of their first year, although breeding may not take place until three or more years later.

Given a very prolonged courtship and intense competition, a suitor must make itself noticed. He or she must also do more, since the potential mate values characteristics that are useful for rearing young. Could showing a good carcass to a suitor be an act of courtship? A raven might find a carcass by sheer luck, so a much more reliable indicator of overall vigor and foraging potential might be flight behavior. The flight displays in unison may be fun, but they also weed out lazy birds and poor flyers. Alternatively, they might be practice flights used to polish up for serious encounters later.

The general explanation for these flights with acrobatic rolls and "rrock" calls is play. Undoubtedly the birds are motivated by immediate enjoyment. But play is not an explanation from an evolutionary perspective; it has a function. I wonder whether the social play of ravens isn't similar to a dance where teenagers get to know each other. Doing the raven "rrock" and roll may be another version of doing the twist and shout.

INDIVIDUALS

JANUARY 23, 1987. A big snowstorm was predicted, so I left Vermont as early yesterday afternoon as possible, hoping to beat it. Of course, it caught me anyway, as I knew it would. But at that point I was already committed. I *had* to go watch the marked birds.

It was supposed to snow some twenty-four inches, and when I was a little more than half way, near the White Mountains in New Hampshire, it was already very difficult to see the road, but, thanks to four-wheel drive, I crept slowly toward Maine. After making it to the hill at night, I donned snowshoes to walk up to camp. It was so cold, and the wind was blowing so hard, that my hands went numb in seconds as I took off my gloves to adjust the showshoe harnesses. Without snowshoes I could not walk at all. It was more satisfying than usual to succeed eventually, and to walk on in a white-out in the night.

In the morning while it is still dark, I dig down to where the baits were left three days ago. Ravens had been back, because two of the fifty-pound meat piles are gone. The birds had also been back in the trap, cleaning up the remains of the meat there.

After considerable trouble I manage to locate and dig up the two other fifty-pound meat piles that had been covered up three snowstorms ago, and that no raven had attempted to dig up, even though dozens of the birds had seen them before they were buried. Here is more evidence that ravens don't find meat by smell, except through the coyotes'. They have to see their food, or else remember where they had already been feeding on it, before they bother to dig.

At 8:00 A.M. one bird flies over the uncovered meat, lands in the trees nearby, and trills intermittently for about half an hour. There is something significant about the trilling and knocking before feeding begins. I see a second bird come, and then I hear several series of knockings. Does the male trill and the female knock to take credit for discovery or to draw attention to themselves in the context of a newly found carcass?

In a few minutes I see one of the birds we've marked, Red 0 (R 0), a known juvenile. It and an unmarked bird are the first ones down. The unmarked one walks up to one bait and then paces back and forth between the two meat piles, picking at one pile and bowing down, fluffing out its head and making strange sounds to other ravens that sit around nearby on the ground. It also makes a series of knocking sounds characteristic of females.

R 0, in contrast, struts around like an adult male, erecting his ears, puffing out his throat, lowering his flank feathers that now look like baggy pants, and drooping his slightly spread wings. There are only three to four other birds around, and none has started to feed. It looks as if they are loitering, not interested in feeding but staying for other reasons. Are they courting?

R 0 is a mere juvenile; but under the influence of another juvenile, R 26, who now appears, he struts with his bill up in the air. The ravens' erect "horns" or "ears" signify power, and once having established it among his peers, R 0 gets down to his main objective: R 26, the young, possibly single female. It is likely that a show of power impresses other ravens, and it may establish the right for courting to proceed. And R 0 exercises this right. He tucks away his horns at intervals, fluffs his head and neck out to the fullest, and bows humbly in front of her, saying, no doubt, sweet stereotyped nothings and signaling his excitement and admiration by blinking with the white nictitating membranes over his brown eyes. She, however, is not ready for his advances. She retreats. He follows and is rebuffed repeatedly. So the "pairs" I have seen until now in the crowds may not have been

mated adults at all, and the courting I saw in my juvenile
cage birds was not an aberration. Juveniles court even in the
wild.

The birds all leave, and it is very quiet for fifty-one minutes. Suddenly twelve arrive at the bait, some of them doing
the jumping-jack maneuver. One walks up and takes a bite,
and they all leave but come back in less than a minute and
leave again. Thirty-six minutes later they finally gather
around the bait with about thirty-five birds feeding simultaneously. Red 0 and Red 26 were not among the first feeders,
although they joined in later.

There are never more than five or six marked birds here
at any one time, while the total number of feeding birds is
about thirty-five to forty. Since we marked forty-one birds
with wing tags (and two with radios) and only 1/8 or 1/7
of the thirty-five or forty birds now here are wing-marked,
the marked birds may come from a total sample of at least
$41 \times 7 = 287$ birds! (The number is higher if it is not a closed
population.) This fits in with the other data, showing birds
coming from several roosts. No wonder hundreds of pounds
of meat were taken within days. In all, throughout the day,
eight different marked juveniles and two marked adults (a
pair) come. No wonder feeding started again very soon after
the Raven Roundup: We caught only the sample of birds
arriving near 7:00 A.M. Most of the others came later and
may have known nothing about the capture. They now nonchalantly walk into the trap, and will teach others that it is
"safe."

Three pairs of adults show up. Pairs are easy to spot because now, close to nesting time, they always stay close together, walking two by two. Both strut with head erect, bill
pointed up and "ear" feathers elevated, while the rest of the
head feathers are sleek. The smaller one with less exaggerated macho features (the female) often follows the first and
preens him on occasion. They frequently touch bills. The
three pairs were W 1 and W 4; a pair where the male had
a crippled right leg (turned-in toes) and his mate had a
white fecal spot on the back; and another pair that had no
distinguishing marks.

In contrast to the adults, *all* of the eight marked juveniles have fluffy round heads, at least after they begin feeding. Their necks are usually pulled in, never raised to elevate the bill. They look meek and submissive. This applies *now* even to R 0, who had looked like an adult courting male when no adults were near. When the adults are *not* here, the non-courting juveniles sleek down their head feathers and look more like crows. I am amazed at all the body and feather posturing because it indicates strong social hierarchies which are likely to be relevant to recruitment.

I did not see R 0 sidle up to R 26 again all day.

JANUARY 24. I put meat into the trap last night, wondering if the ravens will go back inside. They do—even in preference to the newly uncovered pile of meat they eventually fed from in the field yesterday! Old habits are hard to break. End of experiment. I cover the meat in the trap and put about thirty pounds of fresh meat on the pile they fed from yesterday.

Since the birds have been temporarily disturbed, I take the opportunity of being outside the cabin to go check out the other hundred-pound meat pile that has been available since last week on the other side of the mountain. I see a single unmarked raven there. Is it the resident bird? The meat pile is close to an annually used nest.

Today, W 1 and W 4 show up again. But the other two pairs are absent, at least in the afternoon. I see five of the same marked juveniles as yesterday, and also five new ones. Another newcomer is a two-year old B 7, which acts indistinguishably from the juveniles.

The birds stay away for about an hour after I come back to the cabin. All is silent. Then I see a bird fly by giving very deep, resonant quorks. I see one raven chase a marked juvenile (red). More long quorks. Six minutes later there are yells, and about thirty ravens come pouring in and onto the bait in one minute. Two adult birds are there first, and they peck at a number of juveniles. In another few minutes there are over fifty birds, and the adults stop lunging at the others, although they still strut stiffly with drooping wings and erect ears.

W 1 and W 4 come only after the others have already been

feeding an hour or so, as does B 7. Several birds we had not
marked can be easily identified by unique white spots (fecal
markings) on the back. I also see a one-legged bird that was
not here yesterday, but the bird without any tail is not back.
There is clearly a very rapid turnover of identifiable individuals from one hour to the next, which shows that this is
not a coherent "swarm," and this swarm gives one another
solid piece to the puzzle.

I spend a long time watching an unmarked adult pair.
Both have their ears raised continuously, making them easily
distinguishable from the juveniles, but not from each other.
One keeps bowing its head, then holds it motionless while the
other preens it meticulously. Occasionally they touch bills.
A very affectionate couple. They are close to each other at all
times. When a juvenile comes close, it is attacked. But usually the juveniles are ignored, and the couple goes as before,
one feeding, one preening.

At 3:20 P.M. the crowd leaves, after idly perching around
in the middle of the field. One bird with a white marker (an
adult, but I cannot read the number) remains perched on a
poplar near the edge of the field after all the others have left.

Later I suddenly see three birds zooming by in a chase.
The three, who sometimes break up into two, dive and tumble all the way to Hills Pond, come back over the bait, and
plunge out of sight again down the ridge to the west. I hear
the short, agitated "ka-ka-kas" that I have heard many times
during the past two years without knowing the meaning. The
meaning is aggression. I see four times that the chasing individual has the white marker—an adult. One of the two
chasers now comes back and lands again in the poplar near
the bait where an adult perched after all the juveniles left.
It has fully erect "ears" and large "trousers." It makes a
series of long, deep quorks. Yes—the resident pair is defending their meat! These observations confirm it. Another piece
of the puzzle is dropping into place.

JANUARY 25. This is my third day of watching continuously
from dawn till dusk through the spotting scope, to see who
is showing up. I cannot take breaks because I don't want *any*
marked birds to slip by undetected. I again see chases, and

the chasing bird has a white patagial marker, as before. This is fantastic! Consistent observations two days in a row!

R 0 and R 26 are also back. But they come and go independently and pay no attention to each other that I can see. They either never really had a relationship, or they broke up.

Finally, one of the few *unmated* adults shows up. Until now all the adults I had seen were mated pairs, and I had thought there was a specific adult posture. But this single adult looks and behaves and has feather postures indistinguishable from those of submissive juveniles. And the *courting* juvenile looked indistinguishable in feather posture and behavior from the dominant mated adult. So, feather posture is not merely an age characteristic. It provides information, such as status and intent, that is *correlated* with age without having anything that is causally related to age, i.e., increasing age *per se* does not cause birds to change feather posture.

Previously unseen marked birds continue to pass through, and many marked birds who have shown up before do not reappear. I am amazed at how fluid the membership of this feeding crowd is. There are often totally different birds from one day to the next and even from one hour to the next. I would never have discovered this without the marking. Now I can in good conscience discard hypotheses of sharing based on kin or reciprocal altruism, because here at least there is no socially cohesive group.

On the first day of watching, I recognized twelve individuals. On the second day there were eight *new* ones, and four of the twelve of the previous day did not return. Now, on the third day, there are five new ones, and four of the seventeen seen yesterday did not return today. In other words, except for the mated pairs, the birds come and go independently of one another. There is yelling when they are about to go down to feed, which summons whoever happens to be in the area at that time.

After dark each night I go into the woods and up York Hill, where I climb a tall spruce with my radio receiver to get a fix on the two radioed birds. Then I drive to Center Hill near Mount Blue and to several other high points to get cross-fixes of the signal in order to localize it. Today I go out once more to try to get radio-fixes. I got the "beep-beep-beep" of the adult from almost every location I try. As before, I triangulate to the location at the same set of pines. The bird goes each night to within a hundred feet of its old nest. There is magic in *knowing* that the adult stays separate from those going to the communal roosts, which I now know consist of nonbreeders (primarily juveniles), even though they often fly in pairs. I never get a signal from the juvenile; that bird is gone. This is consistent with the color-tagging data. The juveniles are vagrant and do not remain loyal even to a roost.

JANUARY 30. No ravens come all day to the exposed meat! Instead, about forty gather near the apple tree where I had put a meat pile two weeks ago. Only a tiny bit of meat shows above the snow, but they have dug and continue to dig all around it. The fighting is fierce—four agonistic interactions per minute. My new meat pile is eventually discovered, and a bird trills near it. I put even more meat closer to the cabin—and a raven walks up to that, too, and trills. But still no birds feed on the new meat. And no bird trills near the depleted meat pile where many birds are gathered.

In the evening I check out the meat pile on the other side of the mountain. It is snowing hard by this time, and I wallow through snow up to my hips. After much digging I eventually find the remnants of the meat. But it is definitely just remnants. So they have finally eaten it after all. Were the *same* ravens feeding at both sides of the mountain?

JANUARY 31. It has been snowing all night, and the blizzard is still raging. I love walking on this deep snow in my snowshoes with the new snow coming down on me and then driving in the dark along still unplowed back roads.

The blind, which I built last week when I dropped off the meat, is a huge pile of pine branches, now nicely covered with snow. I crawl in and hunch down. In a few minutes a pair of ravens comes, then many single ravens, one after another, in quick succession. The pair are adults, and they act very possessive, jumping and pecking at the juveniles until they are vastly outnumbered. A pair annually nests just half a mile from here.

I see two familiar birds: B 1 and R 15. I saw the latter only

yesterday at the meat on the other side of the mountain by
the cabin. It is clearer and clearer that the birds move freely
between *different* baits, and they do it independently of one
another.

Perhaps each bird keeps tabs on a number of baits, which
explains some of their, to me, mysterious prolonged absences
at any one bait. Maybe they fly around and *compare* different baits, finding one bait overcrowded, or occupied by coyotes, another covered by snow, etc.

As I return to the cabin, the birds are feeding at the new
bait I left there yesterday. (The old has been covered by deep
snow.) As I approach, they all leave. As always, they are
silent; none makes a "warning call" when I get near, as one
might expect in kin groups. But an hour later I hear their
frequent yells, and at 10:00 A.M. they finally come down
again to feed. A whole platoon of juveniles advances to the
bait. After they make contact, they trill and yell, as if to
summon reinforcements. Within seconds there are forty or
so birds. And the feeding begins. It is so common now for me
to hear the juvenile yells before feeding that I often neglect
to put it in my notebook. When something happens every
day, you tend not to notice it anymore. But it is no less significant, perhaps even more so.

Throughout the afternoon an average of four birds out of
forty is marked. I see some of the very familiar ones: R 26,
R 4, and even R 15, who was on the other side of the mountain earlier today, came (B 3 from there did not). There are
even three new marked birds that I have not seen back at all
before this. They must have been feeding elsewhere.

Near noon the birds take off, and when they come back, a
juvenile—a red-marked bird—leads the way. Its head feathers are sleek after the group has started to feed. A pair
joins in. They are very aggressive at first when there are
only a few birds present. But after over twenty juveniles
have gathered, the adults give up attacking others and feed
themselves, instead. Previously, when an adult landed first,
it was immediately followed by juveniles, which it attacked
and managed to hold off for a minute or so. Now I can see
that although the adults *could* be leaders (they fear strange
baits less than the juveniles), there is no advantage in it for

them, because their example will only draw in all the juveniles, who will then think the bait is safe. The juveniles must
balance their fear of the bait with their fear of the adults.
No wonder the adults usually go down to feed after the juveniles are already feeding, at least at those times when they
know there are many juveniles around. They "should" go
down first only when there are few enough juveniles around
so that they can all be chased away.

It snows hard until noon.

FEBRUARY 5. The days are getting longer, but it is still deep
winter. Last night, the northern lights were flickering across
the sky. Tonight the sky is lightly veiled in clouds, and the
quarter moon has a halo around it. It does not shed much
light as I snowshoe up with my gear. I have to make three
more trips, each time carrying about seventy-five pounds of
frozen meat in a burlap bag slung over my shoulder. All of
this is unpaid volunteer work, of course. It is fun. What I do
will never have any major significance in the scheme of
things. So it had *better* be fun.

Finally, near midnight, I'm done with my exertions and
gratefully crawl into a cool but comfortable bed. Alone—
unfortunately. A coyote barks from Gammon Ridge. It sounds
like the dog next door. But out here it seems wild and exotic,
elemental and beautiful. I am paid many times over for my
efforts. But the same things I experience would not be rewards at all if it were not for the efforts I've invested.

FEBRUARY 6. Today I want to finish getting the data for the
graph (see Appendix) I envision showing the individual
marked birds returning and staying away. The graph will
illustrate the fact that the birds at any one bait are not the
same crowd every day, except for the adult mated residents
who come not once but many times each day and stay most
of the day as well.

At the same time, the meat I hauled up is already serving
as bait to capture the second batch of birds to mark. It snowed
lightly in the night, and I must get up before the ravens
come, to brush the snow off the meat I set out last night. I
don't want them to fly by at dawn and see nothing here and

so fly on to their next bait station provisioned by some coyote.

The birds are late. They do not begin to feed until after about 8:15 A.M. As usual, before they feed, there is always a bird who acts as a "yeller." This time I see it. It is perched in the large birch, looking directly at the bait whenever it yells. It looks and sounds very much like one of my captive juveniles when it is begging for food. I think the call is derived from begging. Before the birds fledge, the young beg only from their parents. Later they concentrate on the food the parents bring. My young captive ravens would look at the food and beg from *it* (if I dropped it), as if it would jump up into their mouths. Later they learned to pick it up themselves. The yelling at food has now probably evolved to be an extension of the same behavior, but it serves as a *message* to others, because they have also evolved to recognize what it means. The birds have probably not evolved to remain silent because their calls still help them get food into their *own* mouths: First, it brought their parents so they could be fed, but now it brings its fellows so they can feed themselves. The end result is the same—food in the stomach, as a result of calling motivated by frustration when seeing food and not getting it.

While the juveniles are feeding, a pair of unmarked adults shows up and struts back and forth with bills stuck up in the air. The juveniles crouch, become fluffy, and make appeasement cries. Some even lie on their sides when the adults walk up. One even rolls over onto its back like a puppy in front of its master. The resident adults, W 1 and W 4, also come. No interactions can be seen between them and the other adults, at least on the ground. Perhaps in this huge crowd, anyone and anything can be permitted. Maybe a large crowd is a screen for the adults of one domain to enter and feed in another.

FEBRUARY 7. It snows lightly during the night. I uncover the baits at dawn, but no feeding starts until 9:30 A.M. As usual, W 1 and W 4 are here. And I notice another familiar adult, W 3. As before, it is alone, and it acts and looks like the juveniles. Not only that, it repeatedly jumps away from juveniles, and it is threatened indiscriminately by them. Does it have a low status because it does not have a mate, or does it not have a mate because it has a low status?

Today the weather clears up and becomes very warm. The snow is again spotted with tiny black specks, springtails. Whenever it gets above freezing in the winter, they start to hop, and many hop right off the trees where they apparently spend the winter.

The meat thaws a bit in the sun, and the ravens are again able to tear off chunks, which they haul away to cache. No coyotes have yet fed directly from the baits this winter, perhaps because I have scented the meat with urine. But the woods around here are honeycombed with their tracks. They feed on the ravens' caches. *These*, apparently, they know are safe to eat, probably because they are not contaminated with human scent.

The ravens leave early today, and I have a rare chance to take a walk in the woods. I climb my sentinel spruce to take some pictures of the countryside, and I find a surprise.

Off in the distance, about a mile and a half away in the direction most of the birds come from in the morning, I see ravens soaring and circling. They fly so high that I see only small black specks, and these form into groups of two and three, and then I see the pairs and threes tumbling down head-first in dizzying, careening dives. It looks like tremendous fun. Along the way they call constantly. I hear a lot of the knocking sounds. These are female-specific sounds. Courting.

I am fascinated and watch minutes on end as they circle back up, round and round above Houghton Ledges, then tumble back down. They must be a highly visible and aural target, not just for me, but also for other ravens. Sure enough! I see ravens who have just left the bait fly up and go straight to join the soaring aggregation. The numbers swell quickly. At 2:52 P.M. I see thirteen birds. At 3:07 there are twenty-three with three more on the way toward them. Are they gathering to fly to the roost?

I almost tumble out of my tree, put on my snowshoes, and trot off in their direction. Unfortunately, by the time I get there, there is not a raven in sight. I search the whole area for miles, trotting up to the beech ridges, into the red spruces at the tops of the mountains, and down again into the balsam firs in the valley. Whenever I stop to listen, I hear only the pounding of my heart, from the exertion.

It is another one of those nights when you are grateful to get out of your sweat-soaked clothes and under dry sheets. But first I'm expecting Charlie and his friend Steve from Bowdoin, and two students from the University of Vermont. They are the crew who will help on the second raven roundup tomorrow morning.

FEBRUARY 8. Last night I put the meat two feet forward, *inside t*he trap. But this morning it seems it might as well not be there at all. I'm in my blind at 6:30 A.M., but no birds come till 6:52, and then only two. They answer each other, and one does a lot of knocking. It is the marked resident pair.

The sun sends rays through a few of the tiny holes in the side of the blind. But inside I'm shivering, at first only sporadically, then by eight violently, and it is not very cold today, 10°F. No crowd yet. Did all the vagrants abandon the area, leaving the resident pair? Was that what the soaring aggregation was all about yesterday afternoon? I feel bad for the boys, who have come a long way to see the excitement of a raven roundup and are now waiting impatiently inside the cabin. I make up my mind to hang on a little longer—till nine o'clock. Eight-thirty: I hear ravens in the vicinity. One, a juvenile by its pink mouth, perches in front of me and directly over the bait. It looks down and calls loudly. Its tone and behavior suggest anger! It is not at all happy that the meat is *inside* the cage, where it does not want to go. It flies off.

Ten minutes later a raven lands on my blind, and I hear the heavy wingbeats of others, a soft quork here, a knocking there. They are getting closer. Now one jumps onto the snow. Another joins it. The two walk to the door, look in, fly up and take off, as do those in the trees who have gathered. Ten minutes later they come again, more join in, and they now walk over the threshold. About ten are inside. Should I yank the wire? I think a few more might come in, if I wait a little longer. My shivering is becoming more violent. My eyes can hardly focus, and I have to be careful not to lean onto or touch the walls of the blind, so as not to vibrate it with my shuddering. Steady, steady, I tell myself. Wait. Be patient. Are they in deep enough? How many? No need to worry. Whoosh—in a clatter of wings all are gone from sight. One is spooked, and all take off.

Waiting again. I know they'll be back after they have come this far. And they are back, in another twenty minutes.

Meanwhile, the boys have been taking turns watching through the spotting scope mounted in front of the hole torn in the newspaper covering the window of the cabin. Charlie says that when they finally heard the bang of the door, they all instantly jumped up and ran out to put on their snowshoes.

The catch this time is fourteen birds. One of these is a veteran of the January 19 roundup, when forty-three were captured. And which one is that? Wouldn't you know, it is the one who wears the only number 13. R 13 appeared to be the most cautious of the bunch of marked ones we had seen; it was the very last to make an appearance back in this area so far.

We get all the ravens tagged and released by noon, and we leave just as the sky is darkening and it gets cold again. Before we reach Vermont, we are driving in a blizzard. But the weather doesn't matter anymore. I am elated. Another long weekend of exciting new progress. I can now finish another graph: the age distribution of the feeding "flock." Out of 56 birds it is 6/56, or 10.7 percent, adults, 9/56, or 16.1 percent, subadults, and 41/56, or 73.2 percent, juveniles. There are no data available to show what the age distribution of a random sample of the total raven population would be. I suspect it would be similar to that of other long-lived, late-breeding birds with similar clutch size, such as herring gulls and pinyon jays, where the adults three years old and over make up at least 60 percent of the population. At least 90 percent of the crowd feeders are non breeders. I conclude that the birds who are recruited to a bait are, indeed, a special subset of the population. This is important to know. It is difficult to answer *why* or how the birds congregate unless you can document *who* congregates.

On the way back, we stop in Saint Johnsbury at the diner to mull over and digest events. We wonder how anyone could possibly be interested in the many mundane and often artificial things that seem to absorb so many people, when nature is so *exciting* and so available.

We observe that the excitement is hard won; it requires tremendous energy and persistence. You sometimes need to put in a lot of investment before you can appreciate the unique and interesting.

We then try to justify what we do by trying to make it sound as if it has some "useful" application. But, really, we *do* it because it is fun. Nature is entertainment—the greatest show on earth. And that is not trivial, because what is life, if it isn't fun? I think that the *greatest* contribution we could make would be to help make life more interesting.

SPRING SURPRISES

During the past month I have repeated the experiments of last year, providing simultaneous baits. Only this time I put them over a thirty-three-mile linear spread. It was after a fresh snow, and from the tracks I could see that four of the ten baits were discovered in one day. All were discovered in three days, and then one after another of the thirty-pound baits was eaten by crowds. Coyotes and a fox fed at two of the baits, which delayed their being eaten up, probably because the ravens were held off for a while.

Marked birds, including B 3 and R 34, showed up at the farthest bait, twenty air miles from the marking site. If we consider twenty miles as a radius of the foraging area, the foraging range includes an area of 1,256 square miles. The number of marked versus unmarked birds was the same at the distant bait as at the cabin. Therefore, the same populations of birds are being sampled at both places; there are no separate crowds or gangs at different baits. This means recruitment is not a form of gang protection from other groups. As before, some baits showed immediate recruitment, while others were visited by only one or two birds for several days. The results are finally consistent.

It might be tempting to end the story here: *Voilà*, a hypothesis consistent with all the facts. This is, however, field biology, not a matter of mathematical proofs. In physics, if it works, it is generally considered correct. But evolution has jury-rigged organisms to work in all sorts of ways, and any number of hypotheses are possibilities. I have had a number of consistent hypotheses already, but they became inconsistent with more observations. And so even now I must remain skeptical. Besides, I must convince not only myself. In science the main burden is to provide the evidence that can convince knowledgeable *others*.

MARCH 13, 1987. I do not expect much. There is no reason to. No big decisive experiment has been set up. It is just getting light, near 5:30 A.M., when two ravens arrive. In the gloom I see that at least one of them has a white wing patch, so they are probably the Hills Pond pair that has been here all along. Neither goes near the bait. Strange. Food has been here since last week, and one would have thought these two would go right down to feed. From the signs near the bait, many ravens *have* been feeding here recently.

Others come, but still no feeding. I see chases and hear the familiar short, staccato chase calls. Finally, at 7:40, five birds come down to feed, but they act very nervous. Among them is an unmarked one with erect ears before whom some of the others lie down flat and make appeasement cries. The eared one jabs at a bystander and keeps striding to the bait. There it jabs at some of the others who follow it. A dominant juvenile? All of them eventually feed and then leave.

At 8:55 the birds come again, silently. This time there are two dominant-acting (unmarked) adults. The juveniles prostrate themselves as before. But these adults seem apprehensive. Are they trespassing on W 1 and W 4's domain? And then, suddenly, a real macho-looking adult appears. This one has big trousers and tall erect ears, and from its exaggerated swagger, size, and postures it is obviously a male. Along with him is W 4, the female. It is the Hills Pond pair, the residents. All the other birds now leave, and the pair starts chasing those who are loitering. I see eight aerial chases before 11:07, all in the vicinity of the bait. No birds come down to feed, although I know there are ravens in the immediate vicinity. Until noon (when I leave) I can still hear knocking, quorking, and occasionally a juvenile yell from the forest.

But no birds go near the bait. Is some of the ravens' bizarre "reluctance" to feed not reluctance at all, but the result of the adults' physical repulsions? There have been plenty of hints to prepare me. But now I have an "Aha!" experience that gives me a big piece to fit into the puzzle. The bait has already been fed from, so they cannot fear either the bait or the site. Here, it is the adult, territory-holding birds they fear! And from my aviary experiments I knew the adults are brave and will be followed. Maybe they are here *guarding* the meat not only by chasing others away, but also by refusing to "certify" it for the juveniles.

If one brave juvenile finally goes down by itself, it is almost sure to be beat up on. So it brings reinforcements. Then it has a chance. Is *that* why they always go down together?

If this idea is correct, there may be a dilemma in the recruitment game. If the juveniles are quiet, they may be able to slip in when the territory-holders are away, provided they know the carcass is already safe to feed from. But if they yell, they had better be sure that reinforcements are nearby, because the calls will certainly alert the residents. These few birds were certainly very quiet before and during feeding. Did they know there were no more reinforcements around? I recall that when there were over forty birds here, there was always a lot of yelling just before feeding began, as if they were rallying the troops for a charge. Resident adult birds, of course, should never give food-indicating yells. And I have so far never once heard them do so at any time.

As I have said, you have to set things up properly to do your experiments. What I have "set up" here are the four to five juveniles who hang around, wanting to feed but not having enough clout to do so. *Now* would be the time to get out my loudspeaker and play the juvenile yells that signify the dinner bell. I presume that if I am right, they would now very much like to have company.

I broadcast the yells twice, and each time within a minute not just one, but four very eager birds are streaking out of the woods, circling very low over the cabin, their heads swiveling all around like crazy. (For a fifteen minute control period before each trial I see no birds.) It works like a charm. Now is also the time to try another call as a control. I play two-minute series of the long high quorks, eighteen times in all. Only once does something fly near—within five minutes after the call. It was a *pair* flying over. Is the call perhaps a territorial advertisement? Yes! *This* is the call that I hear the *adults* make on or near the bait. Of course, that's it. The call functions to keep the juveniles *away*.

By noon I am so excited I can hardly stand it. I have to *do* something—run, jump, shout, climb trees. I had long suspected what was going on but had not allowed myself to believe because I have had so many other "believable" hypotheses, and my expectations have for so long been a washout. But I now have an unusually long run of consistent observations. The exciting thing is that very *diverse* observations are now all converging on the same scenario.

I treat myself by escaping from the confinement of the cabin. Before I leave, I smooth out the snow near the bait, so I can tell if birds feed while I am away.

First I drive to bait #10, past the little town of Madrid, some twenty-two miles away. Here, where the tracks say that within the last two days there must have been *at least* forty who overpowered the pair, I again find only the pair, now that I have replenished the bait. They are very busy carrying off huge hunks of the new meat to cache. There are many fresh coyote tracks in the vicinity. I leave my calling card, a dirty shirt under some brush near the bait. Maybe the smell will keep the coyotes off so I can continue to see how the ravens interact.

Thirty-three miles south, near Dixfield, the site where the bait had twice been totally cleaned up in a single day, there is not a single track near the new bait. Obviously, birds do not converge at a bait by flying around at random.

Another interesting thing I saw today was the interaction between ravens and crows. Today there were two pairs of crows, rather than the usual one pair. They were not overtly afraid of the ravens. In fact, two crows edged up to one side of the bait while the ravens fed on the other side. They were only slightly nervous. The ravens, meanwhile, acted as though the crows did not exist. They gave no indication of aggression or intended aggression, either toward one another or toward the crows. One can't help getting the impression that crows are very peaceable birds in comparison to ravens. The ravens share *because* of their aggression! If they were not aggressive, the recruitment/sharing mechanism would probably not have evolved. And I would guess that their specialization of carcasses would also not have evolved because individuals would be fed too infrequently.

How many more goodies can drop into one's lap in one day? I got two more. After I returned from two far-distant baits, I finally took off on a trot on my snowshoes through the woods. It was snowing very lightly. The crust was hard so I didn't break through, and it wasn't even cold. The conditions were ideal for traveling through the woods, over the swamps, up and down the ridges. Mostly I just wanted to romp. But I used the occasion to look for ravens' nests. I had several times heard ravens at dawn down near the brook. Does the resident pair sleep there? If so, perhaps they have their nest there, too.

I hear a raven in the general area, and although I do not see it in the dense forest, I keep walking toward the call. The bird must be moving, because I cover a few miles. Then I hear something different that I recognize from ravens and other corvids—the juvenile begging call of the female at the nest. She has probably just seen the male bringing food, and her yells sound very much like the juvenile yells that also indicate "food here."

The nest looks huge. It is just below the canopy of live branches, far up in a white pine. The snow below is littered with thick, freshly snapped aspen branches. I leave there quickly, not wanting to disturb the birds further. (I later found tufts of nest lining strewn all over the snow, but no tracks. Was the nest destroyed by some of the vagrant juvenile ravens?)

MARCH 14. Another great day, and again the events are unanticipated. I expect the birds from yesterday to be here early, so I get up before dawn. There has been a light dusting of snow overnight, and the temperatures are near 18°F. There is no breeze, and no sound of a raven anywhere, or of a crow or jay.

At 6:02 A.M. a lone raven finally flies over, and when it sees the fresh meat I had uncovered, it almost does a somersault in the air as it backpedals. But that takes only a second. The bird flies on, remaining silent. From its surprise I suspect this is a newcomer, a vagrant.

At 8:03 the woods in the back erupt with the raucous calling of many crows. Twelve minutes later, thirteen of them
are already at the bait. They came as a group, and it certainly does not take *them* long to get started! They seem to
be excited, continuing to caw as they start to cache meat, but
unlike a typical raven crowd, they never shove or jab at one
another. They are feeding close together, almost touching.
I'm glad I have the presence of mind to start my stopwatch,
to get data rather than an impression. It is true: In fifteen
minutes, with five to thirteen birds on or around the bait at
a time, there is not a single agonistic interaction. That seems
to put them leagues above the ravens in amicability. And,
unlike the ravens, not one fluffs out its head feathers at the
bait. I see it only in one crow who is perched quietly on a
tree by itself, all fluffed out. So they *can* do it. They just
don't use it as a gesture at the bait. I see no dominants and
no birds acting subordinate.

Ravens come later, but they stay in the woods nearby, as
if afraid to come down to feed. I cannot count them, but I
suspect there are several. Three times when a raven flies
over, all of the crows fly up as if afraid of the great black bird.

Raven-crow interactions might appear to be between
"stronger" and "weaker" birds. But it is not so simple. In
late April when the crows are incubating eggs, it is quite
another story. Then crows harass ravens vehemently when
they come near their nests. Ravens often fly at relatively low
altitudes, searching the ground for food, and soon enough a
crow will come and fly far above the raven and dive down
onto it. The raven is obliged to take defensive measures by
flipping over to expose feet and bill. And as the crow swoops
on by, the raven immediately rights itself, and the crow flies
up for another swoop. As the raven hastily leaves the crow's
territory, it soon enters that of another, and is harassed in
relay as it flies over the countryside looking for bird's eggs
and other spring delectables.

The ravens who are nearby sound agitated, but they do
not come down all day to feed. They are certainly not afraid
of the crows. And if they were afraid of the bait, the crows
have proved that it is safe. But the crows do not lure them
down. Would another raven? At 10:00 I put my stuffed

raven near the bait. It makes no difference as far as I can tell. The ravens continue to fly over and perch nearby, but they do not go near the bait. Do they not go down this morning because they still fear the adults who chased them yesterday, or are they satiated? Maybe the adults are nearby, and I do not see them or do not distinguish them.

At one o'clock I decide that little is likely to happen now, so I take another walk in the woods. I find the place where a raven walked in the fresh snow and tore off the inner bark from a dead ash tree. Twenty minutes later I have located the nest. It is also in a white pine tree, also new, and within fifteen yards of the tree that last year held the nest from which I took young for my aviary studies. A pair had here enjoyed exclusive rights to a bait I put out, while other baits had been overrun by crowds. Since the bird is building the nest lining now, I expect she will start laying in three to four days. Give her another four to five days to lay four to five eggs, and twenty-two days to incubate them, and the chicks should hatch a little over a month from now, in mid-April. There will still be deep snow on the ground.

I get back to the cabin at 3:30 P.M., dog-tired from miles of trekking on snowshoes. But I'm happy and excited to have found another nest. I will mark the young in May, to try to confirm published reports that locally raised young disperse by winter.

It is starting to snow when I get back, and I have just time to check the tracks at the stuffed raven and the bait. Not much has happened. One raven landed in the snow and flew off after taking two to three hops.

I saw a lot of soaring activity from the ravens here today. I'd never before seen it so early in the day. Is it because they were loafing, waiting to start to feed?

MARCH 15. The third day, and five or so ravens are *still* hanging around for large portions of the day and still not venturing to the bait, although some now land on the snow in the clearing. The juvenile R 26 is among them, seemingly quite friendly with another one (unmarked) who assumes an adult strutting posture (i.e., a dominant bird). Isn't R 26 the same bird who spurned the advances of R 0 a month ago? Yes!

I hear what sound like aggressive interactions from the nearby woods. Is a pair of adults still keeping them at bay? And are they trying to get more recruits? Is that what the soaring yesterday was all about—a maneuver to provide a target to bring others back into the area?

MARCH 21. It is the official first day of spring. On this date the very first red-winged blackbirds come back to Burlington, Vermont, and if one is lucky, one hears the first spring bird song—the robin's. The day it comes back is also the day it sings to establish its territorial boundaries. But right here at my study area in Maine, the snow is still deep, and it is snowing again.

The big meat pile that was nearly untouched all last week has been totally cleaned up by ravens over the last several days. The whole field in front of the cabin is trodden down with their tracks, and as I come up the trail in the late afternoon, near dusk, six ravens fly off. Only one of them has a tag, a red one. I put out forty pounds of fresh meat.

MARCH 22. Up to twenty ravens are here all morning, but only one is marked. Familiar old R 26. This juvenile and others are continually caching the soft, unfrozen meat. Clearly most of the marked birds have left, and most of the birds in this area are now new. Why has R 26 stayed? It is clearly a very dominant bird, amongst the juveniles.

One interesting thing I notice is that for over an hour before and after the birds (with R 26 among them) start to feed this morning, there is not one yell and not one trill. The birds are quiet, and none of them has a fluffy head. Is that because today there are no resident adults nearby?

Near 7:00 A.M., however, I see an adult strutting among them, pecking at birds right and left. Now for the first time I also see birds fluff out as the adult approaches. There is no doubt any more that the fuzzy tucked-in heads signify submission and low status—an expression of respect perhaps in the presence of the lordly territorial adults.

Meanwhile I have brought with me a new tape of juvenile yells, and after removing the bait and seeing no ravens nearby, I play it six times. Within one minute birds come and circle the cabin, obviously very interested. The number of birds that come (usually within ten seconds) during the six consecutive trials is 4, 2, 2, 1, 1, 0. It looks as if they catch on quickly to false advertisements. For the last three trials I uncover the food, but the birds still do not land; they are interested in *company* with which to feed. They already know about the bait.

The results today again confirm what I had found out before, that the yells recruit, and that the adults are dominant. But it is always a thrill to see it again, in a slightly different context.

APRIL 1. I heard the first lone spring peeper down in the pond last night. In a week or two the frog chorus will be deafening.

Here in Vermont the crocuses are blooming. The crows are building their nests, and some of the red maples are starting to bloom. Most of the snow at low elevations is gone. Canada geese are migrating north. And in Maine the ravens have been incubating their eggs for a week or two.

In Maine there are also big snowstorms followed by record floods. Torrential warm rains following snowstorms have been quickly melting the deep snow. Rivers are flooded, bridges torn out, and roads are under water.

APRIL 17. 6:30 A.M. A raven with erect ears is already giving deep rasping quorks from the top of a red maple tree. The carcass has been noticed. Another raven, which I can identify by its split scraggly tail, flies over and silently lands in the woods. The first bird flies up, and I soon hear lots of defensive cries. Split-tail gives its appeasement gesture by crouching down and vibrating its spread tail in the female copulatory posture (given by either sex). The eared raven has no mercy, it moves up to split-tail again, who gives more appeasement cries and leaves, to be chased to the other side of the hill. (The same crouching copulatory posture with vibrating tail was offered to me by my tame dominant male after I had been absent for a few days. It was also given by the subordinate female to the male after he repeatedly tried to chase her from good food she wanted to eat.)

At 7:00 a raven comes silently, goes quickly down to feed, and wolfishly tears off chunks of meat, which it takes away to cache. It is back again quickly, tearing off more meat. It makes deep territorial quorks while still on the meat, but only when two other ravens appear flying over. They land in a nearby tree, and the caller, the "owner" of the meat, flies up to them. One of the newcomers flies off. The other remains and goes through the appeasement routine of crouching and tail vibration. It is of no avail. The aggressor erects its ears, lowers it flank ("trouser") feathers, stands erect, fluffs out throat feathers, and advances again. The appeaser flies a short way off after giving a protest call. This pattern is repeated eight times before the appeaser leaves the premises. Shortly after this a red-tailed hawk comes by. Four ravens scatter out of the woods. Where had they been hiding? I did not know they were there until now.

Twenty minutes later a single raven flies by briefly and makes the long, undulating territorial quorks. At 11:20, long after it has left, split-tail finally comes by again, alone. No other raven is near, and *now* it feeds. After this I also leave, being very well satisfied.

AT THE NEST

A FRIEND WHO KNEW about my interest in ravens told me that one spring some years ago he had heard ravens at a remote granite quarry near Adamant, Vermont, and in the bottom of the quarry there had been a pile of sticks, the likely remnants of a fallen raven nest. Adamant is less than an hour's drive from my home, and so I went there to investigate in early April.

It has probably been a century or more since granite was mined at the site. Now there is a seemingly natural cliff surrounded by woods, with a jumble of boulders around a water-filled hole at the bottom. From the cliff you look out over a marvelous cattail marsh with a huge beaver lodge in the middle. Beyond there is a bucolic farmstead among the forest on the hills across the marsh.

As soon as I walked into the woods near the cliff, one of the ravens made the oft-repeated rapid "rrack, rack, rack" calls and flew around the surrounding forest with rapid, shallow wingbeats with its head feathers fluffed out. This is the behavior that signifies mild agitation when someone comes near a nest.

It is always a thrill for me to find a raven nest. Birds are closely bound to their nests, and when you find one you have a way of getting close to the birds. This nest was, like almost all raven nests on cliffs, tucked under an overhang. It was also some twenty feet above jagged boulders. What made it special was that I could look down into it without climbing up from the jutting edge at the top on one side of the cliff. There were five eggs in the nest.

I immediately set about collecting dead trees and live spruce branches from the surrounding forest and built a blind at the top of the ledge. When it was finished, I had a fantastic view of the nest from about thirty feet away. (I would later find that it was close enough for a raven to be seen full-frame with a 400-millimeter lens. But the natural amphitheater of the cliff and the large overhang over the nest made it too dark for decent photography.) I left to let the birds get used to what, to them, would be a new pile of brush, but I was anxious to be in it.

APRIL 18. Woods filled with early spring flowers are immediately to my back. The bog with frogs is to my front. I am in a blind that is thatched together out of dead trees and spruce branches, and I am watching the raven nest on the cliff, which faces the marsh.

Spring seems scarcely here, but the wood frogs are almost done breeding. A few are still calling like frenetic ducks down below me. Two days ago we had a big snowstorm. I hope it will be the last one this season. The ice is off most but not all of the ponds and lakes, so spring is definitely on the way. The frogs can't be wrong. Also, as I look over the hardwood forest on the hills opposite the beaver pond, I see a faint tinge of red. The red maple buds must be swelling, and their color contrasts with the mouse gray of the twigs and branches. Today the sun shines through the branches, melting the new snow and revealing the brown foliage on the forest floor that is now adorned with bloodroot's brilliant white flowers and saffron yellow anthers. The pink spring beauties are also out. And there are a few scattered patches of dainty hepatica, some white, some pink, some bright blue.

There are four young ravens in the nest, ranging in age from about five to eight days. (Commonly one egg does not hatch.) The oldest is just beginning to extrude the tips of pin feathers from wing stubs. But they all still have white-edged bills and naked pink skin showing the "four o'clock shadow" of dark feather growth that will soon erupt in tiny quills. Their eyes are closed most of the time, but they fidget a lot, as if not sleeping very soundly. One of the youngsters almost continuously holds its bright pink mouth open and its eyes closed. In addition to fidgeting and flopping their heads onto one another and on the edge of the nest, one occasionally stretches its neck up and gapes, as if expecting food. Maybe it is dreaming. When they are not moving around, they huddle together with necks laid over one another, and the pile heaves from the collective breathing.

The nest, like other raven nests I have seen in New England, is built on a foundation of poplar twigs, freshly broken off the trees. You can see the fresh yellowish wood at the end of every twig, and some of the twigs are three quarters of an inch thick. It takes a big bird to break off fresh branches this size. The deep nest cup is lined with strips of bark and deer fur. When I first peered into this nest a little over two weeks ago, I saw five greenish eggs, variously marked with black and gray. Like most nests on cliffs, this one is protected from the elements, such as the recent snowstorm, by the overhanging ledge.

I now hear the swooshing heavy wingbeats of a large bird—and a second later a raven lands directly on the edge of the nest. The young hardly react. This surprises me. I had expected that they would pop up like jack-in-the-boxes and beg vehemently. The adult's feathers are a dark metallic blue, very smooth and shiny, and they glisten like a giant dewdrop in the sun. This is not at all the common image of a shaggy black raven. It cocks its head to the side and examines the young briefly. Then it delicately picks up piece after piece of the white feces that the young have left on the edge of the nest. Each piece is swallowed.

The adult now does a strange thing. A stream of water is running down the side of the cliff onto the edge of the nest, and the nest lining is becoming soaked. The bird busies itself for twenty minutes poking into the nest, as if letting out the water. Then it stands still and erect over the young, who poke out from underneath its breast feathers. Even now they do not beg. The raven fluffs out its breast feathers and gently settles down to brood the young. A half hour later it abruptly stands up, launches itself from the nest, and flies off across the bog on slow gliding wingbeats.

Less than half an hour later, this raven or its mate arrives again. Its throat is expanded. As before, it lands on the nest, and there is still no reaction from the young—not, at least, until the parent makes one short grunting sound. Instantly the four erupt in hoarse cries as their necks shoot up, and the four bright red open mouths weave on long scrawny necks like red poppies in a breeze. The raven sticks its bill into one or two of the gaping maws, regurgitates food, and in mere seconds leaves the nest, flying off into the distance.

MAY 6. I'm again in my pygmy hut of evergreen boughs at the edge of the cliff by the raven's nest. The four young are already clothed in dull black feathers. Two of them look alertly with their light blue eyes over the edge of the nest. (The adults have dark brown eyes.) Two others have their eyes closed. All are tucked snugly down into the nest cup. They soon fidget and jostle but remain silent. Below, where the boulders clothed in luminous green moss are strewn among the remaining ice and snow at the edge of the cliff, the tiny brown winter wren and the northern water thrush sing their loud, liquid yet very different songs that echo off the walls. Out in the bog the spring peepers sing, and within their constant din I hear the loud reverberating pile driver "ker-thunks" of a bittern. For continuous background music there is the steady yodeling and cackling of the red-winged blackbirds, the trilling of the swamp sparrow, and the melodies of the white-throated and song sparrows.

There is no raven in sight, and my ears continue to become more and more attuned to the surroundings. In the woods close by, I notice the different pitch and frequency of hairy and downy woodpeckers drumming. I hear chickadees making their plaintive two-note mating calls, the nasal twang of the red-breasted nuthatch, and the high reedy repetitive "tseets" of the black-and-white warbler. There is the leisurely refrain of a solitary vireo and the loud rapid cascade of an ovenbird's song. Finally, off in the distance, there is the occasional cooing of a mourning dove, the excited staccato of a flicker, the song of a robin, and the cawing of a crow. In all I detect twenty-one different bird voices in just ten minutes. In another two weeks there will probably be more than double that number. What strikes me as remarkable is how totally different all the bird voices are. That is undoubtedly not an evolutionary accident. The more different one is from another, the easier it is to pick out from the cacophony. And that is undoubtedly as true for those to whom these sounds are meant as it is to me, an idle spectator. Each song is a signal. As such, it must be unique.

The young ravens seem to be restless now. One backs its hind end up and over the edge of the nest, wiggles the stump of its tail, and voids a white liquid stream over the edge. Unfortunately most of the young do not have the advantage of a readily available nest edge. The overhanging cliff prevents that. Another yawns, exposing its bright pink maw. Another retches and regurgitates a light-colored pellet, which rolls down into the nest cup.

A mourning-cloak butterfly flies near and lands on the rocks just below the nest, lapping up liquids undoubtedly rich in minerals. It flies up to the nest, where one youngster snaps at it but misses. A bumblebee queen drones sonorously around my blind, searching for a mouse nest in which to start her colony.

One of the raven parents finally returns and lands on a ledge just below the nest. From there it hops up to the edge of the nest. The young gape and beg in their imploring, rasping voices and are quickly fed. The parent's vocal signal to get them to gape is no longer needed or given. In a second or two, all is quiet again, and the young plop back down like so many limp rags, stretching their necks out over the nest edge.

The adult (it is the female, as I will determine shortly) stays. She bends down and with her great thick bill gently nibbles among the feathers on the top of the head of one of the young. She continues the treatment to the base of the young bird's bill, then she tenderly preens the tiny feathers around the eye. The youngster seems to find this treatment pleasurable because it stops fidgeting, closes its eyes, and makes soft low comfort sounds. After ten minutes she reaches over it and starts on the next, then preens the other two as methodically. All told, she works on them for over half an hour, then I hear wingbeats. She stops abruptly and leaves the nest. Her mate has returned.

He flies directly onto the nest, and they all gape and beg.

In two seconds he has unloaded his swollen gular pouch. Now he starts to preen one of the young, but his heart does not seem to be in it; he stops after about ten seconds, gazing at them with head cocked sideways. His alert gaze shifts from one to the other.

I presume that this is the male, because it is a larger bird with a longer and thicker bill. The smooth silky feathers on the sides of his head look as if they are combed back, and they produce a ridge on the back where they come together. His posture is different, too. (At least on this occasion.) The female always has her wings tucked in, making her look thin and trim. But he droops his wings slightly and spreads them apart at the shoulders. He gives his young one more glance, then pushes off the nest and dives from the cliff in one easy motion, landing on a nearby poplar.

The female now comes back to the nest and immediately resumes her preening. This time she works uninterruptedly for twenty-eight minutes. Her feathers look gun-barrel blue. Every feather is distinctly outlined, in contrast to the youngsters', which are dull and fuzzy and blend together. The male, meanwhile, returns from the poplar and perches on a sunny ledge near the nest. There he looks all black and white as the sunlight mirrors off his smooth feathers. He soon begins calling in a variety of sounds, and his mate leaves the nest and flies over to perch close beside him.

Together the two make soft, tender squeaking and murmuring sounds, as one closes its eyes, dips its head, and presents its fluffed-out nape to the other to be solicitously preened. Now one slowly reaches over and grasps the bill of the other, and so they stand side by side, silently holding bills, for minutes at a time. After half an hour of tender exchanges (it is difficult for me not to anthropomorphize), the female makes a few short quick quorks and flies back to the young yet a third time. She preens them all again, this time for thirty-two minutes.

Meanwhile, the male leaves his perch on the ledge, circles above the cliff making several series of short rapid calls, and departs to the southwest, the same direction in which they have been consistently flying to forage. A crow calls briefly in that direction, signaling that the raven has passed. The female seems to ignore her mate's call, continuing her preening of the young. A broad-winged hawk soars overhead and calls, but she pays no obvious attention to it.

MAY 7. The deciduous forest now shows pretty patches of red and green through the background colors of gray and brown. The red maples are blooming, and some of the poplars have pea green leaves bursting from their buds. Here at the cabin in Maine, tree swallows are investigating one of my bird boxes. The early warblers—Nashville and black-and-white—are singing vigorously. I heard a single song note of a black-throated green warbler. The white-throated sparrows sing in the morning and even after dark. I hear a purple finch. A pair of evening grosbeaks stops briefly on the birch beside the cabin. In the distance I hear the occasional blue jay, cowbird, golden-crowned kinglet, and woodpecker. In the evening the woodcock displays in the crude field in front of the cabin, the first hermit thrushes sing from the woods beyond. Large blue-black tachinid flies chase each other after basking on the logs of the cabin. The pesky black flies have not yet emerged. I give them less than a week now.

One can't help but be impressed by the sudden activity of all the birds that have been absent for so long. I marvel at the miracle of their return. But my objective back in Maine this weekend is ravens. If the young at the Adamant, Vermont, nest are any indication, it is time to tag the young at the three local nests near my Maine study site. The more local birds I have identified, the more I will be able to learn next winter.

The nest up in the pine at the Grahams' Farm is the first one on the agenda. Last year this was where I found the single fledgling that survived the unusually heavy late snowstorms. It was marked (Y 1) and was seen by several observers throughout the area in the following summer. This year the nest is about two hundred feet from where it was last year, but only about fifty feet from where it was two years ago. (By contrast, when the nest by Hills Pond blew down in the fall, it was rebuilt in the spring in exactly the same place on the same tree. A year later it was a quarter of a mile distant, on the other side of the pond.) This year

the nest is also at least a hundred feet up, near the very top
of a white pine growing tall in a stand among many others.
There are no limbs on the tree for about seventy-five feet,
and I do not relish having to climb this monster because I
have to use climbing irons over my shoes. I feel ill at ease
probing with my foot to get a spike-hold in the brittle bark,
which will hold my weight only if the angle is just right. My
tendency is to hug the tree, and this causes the spikes to slip
and my feet to flop around, especially as I get higher up and
more and more tired.

One of the pair makes the deep territorial quorks even before I start to climb, and I immediately hear its mate answer
with the same calls. But as it did last year, the second bird
quickly leaves the scene entirely, while the first becomes
ever more frantic and brave as I get higher. It now makes
high-pitched staccato calls, and, again like last year, lands
on branches all around me, coming to within about fifteen
feet as I approach the nest. It vents its anger by hammering
branches and clipping off twigs. As I watch those twigs fall
from the top of the pine, which is swaying in the wind, the
ground looks very far away. I am tense and shaking from
exertion and fear. But I've made it to the top.

As I maneuver myself around underneath the massive
platform of sticks in order to get over an edge, the adult that
has stayed goes into paroxysms of angry calls. Now the
young in the nest are raising the chant, too. They make long,
rasping, growling calls like so many snarling dogs. As I
finally peer over the nest edge, all four nearly grown youngsters stand up and crowd to the farther side and raise their
calls to a new crescendo. I now pull on the string attached
to my belt and lift the tote bag from the ground. Then I
grab one hefty struggling youngster after the other and put
each into the dark bag. Surprisingly they do not bite, and
they are temporarily quiet as I lower them down to the
ground with the string.

After I have made my way back down to the ground, I'm
reminded that the exertion was more than I had realized.
For about ten minutes my hands are so weak that I can
hardly hold a pencil. But while I was on the tree and needed
them the most, they worked wonders.

With the crew of helpers, Alice, Denise, and Jesse, who came to witness the climb, we eventually manage to put on the yellow plastic wing tags. It is a big struggle. These birds are not cooperative. Every motion that any of us makes is a threat to them. They have met the enemy, and it is humans. One parent told them so, and their experience so far has confirmed it.

I was physically unable to repeat the climb to put the young back. Instead, I built an imitation raven nest in a nearby tree, to which I *could* climb, and I put the young into this nest. I knew the adults would care for their ready-to-fledge young no matter where they were. And, indeed, as we were leaving, one of the adults had already flown over to them to feed them.

The second nest was quite a contrast. I found it only two weeks ago, within a mile or so of where I had previously seen ravens on several occasions. Once you know where a raven pair lives, you can find its nest. I doubt if any other bird nest is easier to find, after you begin to think like a raven. This one was also in a tall white pine, but the tree had limbs, for which I was extremely grateful.

As at the first nest, both parents were away when we arrived. But I was barely at the top of the tree when one flew over with a bulging throatful of food for the young. It made a few calls, and its mate was summoned almost instantly. They made their territorial quorks, but I did not feel that they were highly excited. Both birds stayed well away from us and the nest. Will these young also be less agitated? After I've climbed almost up to a nest, I take the added precaution of talking to the young softly, as if I were speaking to a baby, in order to calm them before I even peer into the nest.

This time, as I first peek over the nest rim, the young (the same age as those from the previous nest) duck down, showing a little fright. But there is no panic. None jumps up and growls. I keep talking to them, and eventually lift all four out, one by one, and put them into the bag. They do not struggle. The parents keep calling, but only from a distance. I hear none of the harsh angry calls that the previous pair gave.

After I lower the brood to the ground and descend leisurely, we lift the first youngster out of the bag. It makes a little, plaintive noise. Is that a hesitant beg for food? Incredibly, the youngster fluffs, shakes, and then nonchalantly starts to preen itself! There is nothing more captivating than a fluffy young crow or raven that is at ease and friendly, and we only reluctantly interrupt its toilette to put on the leg bands and wing tag. No sooner do we get it done and set the bird down, than it continues as before, as if nothing had happened. And so it goes from the first to the fourth, until we have all four fully feathered raven youngsters sitting among us four humans as if it were the most natural thing in the world. Everyone is instantly in love with these totally trusting birds, and we marvel at the contrast with the last batch, which we were glad to be done with.

I hate to break up our little party, but I have to. I put them all back into the bag, climb the tree, pull up the bag and set them back again into the nest. They settle right in and continue their preening, and by the time we have gone some hundred yards away, the parents have rejoined them at the nest.

The third nest is on a cliff, not far from nearby Dixfield, and I am lucky to be able to get to it at all. One previous year I had scaled this nest by propping a slender tree against the cliff and shimmying up the pole. This year I have manpower enough to bring a ladder through the woods and up the mountain.

Both parents sound the alarm as we get near. As usual, we first hear the deep territorial quorks, which are replaced by the higher-pitched, excited repetitive calls, given with rapid wingbeats and fluffed heads, when we get close to the nest. In general, the parents' vigor of response is halfway between that at the other two nests.

I talk to the young in soft tones and handle them with care, but they continue to be afraid, ducking their heads deep down between their shoulders. We work at the base of the cliff, so that the parents can see us and their young. The adults' calls gradually become muted, but as we work quietly, the young continue to sit ducked down on the ground around us. We now hear the "thunk" calls, which I had previously thought were the all-clear signal, saying that everything is all right. Is one bird reassuring the other? We heard even more thunk calls when we finally left, and I now recall having heard the same calls when leaving other nests.

MAY 15. This promises to be a beautiful sunny day, and so I have again come to watch the cliff nest at Adamant, Vermont. Now, at 8:00 A.M., the fog still veils most of the bog, but it is burning off fast. Opposite, the sugar maples have turned from gray to yellow; they are festooned in tassels. Meanwhile, in the marsh one or two male Wilson snipes are cackling from some grass tussocks, and another is winnowing high up in the air. The bittern still makes his pile-driver noise, and the red-winged blackbirds continue their steady concert of yodeling and cackling. All of the other birds I heard the last time are also singing. The winter wren is again close to the cliff, his voice reverberating like an opera singer's among the rock walls. Above me now is a new voice—the melodious, slow whistle of the rose-breasted grosbeak. It is a beautiful song. A pair of chickadees close by now makes soft soothing comfort sounds; the pair is getting ready to build its nest.

Off in the distance I hear a blue jay convention. This is the time of year when these otherwise solitary birds sometimes gather, presumably to socialize and, if unmated, to find mates. I have seen up to thirty convene at a time, but never at food in the winter. So they *can* at least aggregate, but they choose not to at good food.

The raven young are now even more jammed in under the inward-sloping ledge. Nevertheless, one youngster has left the cup of the nest and is perched on the rim. It is continually active. It stands tall and stretches down the right wing, retracts it, stretches the other wing. Then it ducks down and stretches both wings by lifting them. It preens its tail, then one wing, then the other. Now it picks at the twigs of the nest, now at the bill of a nestmate. Its three nestmates are nearly comatose in comparison. But one of them awakens for half an hour and appears to try desperately (and unsuccessfully) to clamber over the backs of the other two to get onto the nest rim. Is something wrong in the nest?

The raven pair is back by 9:42, and within two minutes they have made three nest visits. As they fly up to the nest, the young make their plaintive rasping calls while opening their mouths, and the parents drop in white material from their extended throats and quickly depart. The whole transaction takes about one second. No preening of the young today. And they really need it now! Their feathers are badly soiled.

The fog burns off. The snipes are silent. Occasionally a lone spring peeper still sings, to be quickly joined by a chorus of several others. There are other frogs as well. You hear an occasional green frog croaking; it is joined by more of its kind.

The raven pair has been idling in the vicinity of the nest, but at 10:11 one makes several series of four successive rapid, high calls. The second bird flies over to the caller. They both make low grunting sounds and then leave.

Twenty-two minutes later both are back, but only one of them goes to feed the young at the nest. This bird gives the young what it has brought, then picks a piece of feces out of the nest and, rather than consume the feces itself as it would normally do, drops it into the gaping mouth of one of the nestling which swallows without hesitation. (Later, as the young get bigger, and process considerably greater volumes of food, the adults neither eat nor remove the feces.)

Another half hour passes. It seems like an instant, as I continue to be entertained by the singing of the many birds and the antics of the nestlings. The one on the nest rim is still preening as before. An adult approaches. I hear its wingbeats as it lands nearby. The two runts in the nest awaken and make their begging calls. The adult flies to the nest, feeds them, and quickly departs to land on the nearby poplar. Here it makes several series of quorks. The quorks of each series are identical. But the calls are entirely different. I distinguish at least five. Some are different from those I've heard at other nests. Others sound identical. The bird flies down to feed the young, and there are two more feedings within one minute. Since I see no food in the bill or throat as the bird leaves the nest, I conclude that it regurgitates more each time it leaves and then returns to unload the bolus of food that has just come up.

After four feedings the nestlings are quiet, and the parents settle down beside each other on a top limb of their favorite poplar near the cliff. They nuzzle each other, and make soft grunting and squeaking noises. After about five minutes, one of them makes a series of soft nasal quorks and abruptly leaves. The other stays and preens itself.

It preens for over half an hour. Then it checks in all directions for several minutes until its head finally comes to rest. At 11:52 it suddenly jumps up and takes off from the tree. It flies around, lazily circling above the cliff before departing, making numerous series of at least four different kinds of quorks.

MAY 22. In the past week we have had torrential rains and warm weather. The forest has suddenly leafed out in various hues of bright green. Apple trees are clothed in pink blossoms, and the hay fields are under a yellow carpet of dandelions. I stop at a small trout stream flowing through a yellow meadow and walk to a grove of pine trees that should be an ideal place for crows' nests. But before I even get to the pines, I find a red-winged blackbird building her nest in a clump of newly erupting grass. Close-by, a kingbird flies up in a fork of a dead elm tree with a billful of dried grass. The first crow's nest I find is empty. Strange. Crows should have small young by now. The second nest I find probably has the answer to the mystery: It was first a dinner table and is now a sleeping platform for a raccoon. The animal is snoozing with closed eyes, its head draped over the side.

The Adamant ravens are not near when I get to their nest, and when I look into it, I see a dreary sight. Since the rains, there has now been more water dripping from the cracks of the ledge, and the nest is now a soggy quagmire. The cliff may have protected the birds from snow and from predators, but its location has been less than ideal. Pushed down into the nest's quagmire are three dead young. The fourth is, as before, perched on the nest edge. It preens, stretches, vigorously exercises its wings, and seems healthy. It even hops onto the dead and then off the nest onto the ledge. It explores the ledge, then hops back onto the nest rim.

A raven quorks nearby. The young bird lifts its wings, vibrates them, and calls for food. In seconds the female lands at the nest and regurgitates a bolus of food into the begging red maw. The youngster is immediately quiet again; it was not very hungry. The adult stays to pick several maggots from the dead young, catches a blowfly with a loud snap of its bill, and then spends fifteen minutes quietly perched on the nest edge, staring down at the dead young. Finally she flies off to join her mate on their favorite perch on the poplar in front of the cave.

The two perch close together for several minutes, making their usual comfort sounds, but she soon interrupts the intimacies to fly back to the cliff. This time she lands on a ledge under a second overhang that is about ten feet above and to the right of the nest. I had myself just been looking at it, wondering why the birds had not chosen *this* site over the other one for their nest. Perhaps this ledge slopes too steeply to anchor a nest, but I now notice four poplar twigs on it. She picks up one of them and plays with it for over five minutes, as though going through the motions of incorporating the stick into an imaginary nest. Over and over again she presses a stick down on the ledge and vibrates it with her bill as if trying to push it in. Then she flies down to the ground where she picks up three more twigs in her bill and flies off with them. The whole exercise is a waste of time, because I know she will not, indeed cannot, renest this season. She has a youngster who will remain dependent on its parents until late July. It is too late now to nest and too early to leave the young one. I wonder whether or not the raven is vaguely aware that the nest site she had chosen is a poor one, whether she is now acting out a vague knowledge of what might have been a better choice. Next year may tell.*

* When I returned to the site on February 19 the following winter, I found the ravens' fresh tracks and freshly broken-off twigs on the snow near the top of the cliff. Several new twigs had been placed on the ledge, but nest building had not yet begun. On March 10, when I returned again, the old nest had been completely rebuilt. It was fully lined and ready for new eggs. (Six eggs were laid and five young hatched by April second.) However, although the birds had refurbished the old nest, they had clearly made a valiant, if not foolish, attempt to nest on the sloping ledge. Directly underneath it was a huge pile of fresh nest foundation sticks—1,375 to be exact—enough for at least four nests. Apparently, the birds had very much wanted to use that site, but they were slow to learn that the branches would not anchor there.

RAVEN CALLS

I N AN ARTICLE for *International Wildlife* in 1984, Fred Bruemmer describes seeing two ravens in Alaska which were croaking hoarsely while flying across the wintry tundra. Bruemmer turned to his companion, an old Inuk, asking: "What did they say?" The Inuk smiled: "Tulugak (the wise raven) say *'Tuktu tavani! Tuktu tavani!'* ('The caribou are there! The caribou are there!')"

Ever since, and presumably even before the pair of ravens, Hugin and Munin, perched on Odin's shoulders to tell him the news of the world, the raven's varied utterances have been of great interest to humans. A bird who knows all, and could tell all, is to be listened to, although, as Edgar Allan Poe implies, the bird is not always willing to divulge its secrets. Nevertheless, there has never been a lack of people who claimed to understand him. Ornithologist Thomas Nuttall remarked in 1903 that "all the actions of this sombre bird, all the circumstances of its flight, and all the different intonations of its discordant voice, of which no less than sixty-four were remarked, had each of them an appropriate significance; and there were never wanting imposters to procure this pretended intelligence, nor people simple enough to credit it."

There is probably nothing about the raven, *Corvus corax* (the Latin name comes from the Greek *Korax*, a croaker), that has been more commented on, studied, and written about than its voice. But I'm convinced that there is nothing that we know less about. Indeed, what we know is minuscule at best. Trying to integrate the incredible calls I have heard the Maine ravens give with the scientific literature has been a singularly frustrating experience. Some of the studies have involved the objective use of sonograms, but even these have helped little. Eleanor D. Brown discovered in 1985 that different social groups of crows have different song repertoires. Could it be that there are strong local raven dialects? Or do the birds have only a few standard calls and improvise the rest of the time? It is one thing to recognize the different vocalizations, still another to decipher their meaning. So far we have not made much progress even on the first.

One of the first people to make sonograms of raven calls and study their meaning was the German zoologist Eberhard Gwinner who worked with several groups of tame captive birds in Germany. He acknowledges that his ravens had a very large repertoire of calls that were difficult to classify, in part because the birds mimicked sounds and had strong individual characteristics. His publication provides sonograms for some of the more distinctive ones. These include a short "rapp" or "krapp" which he identified as the *Flugruf* ("flight call"). Another is the loud *rüh* call or *Standortruf* ("place-indicating call") given by hungry young to summon the adults and by the incubating female to summon her mate. The loud "kra" calls were given while ravens were defending food. In addition, the kra, which Gwinner considers the typical raven call, is used in situations where there is real or imagined threat. This is probably the call often written about in English literature as the "pruk," "kruk," "quork," or "croak." The "gro" calls, or contact calls, are given only when another bird is already near. And then there are soft whining calls (*Winsel-laute*), a variation of the gro calls. These are given in intimacy between birds friendly with each other. Gwinner also describes several calls used in self-assertive displays, and a knocking mechanical sound he thought was the mimicking of the white stork's call, although it was given primarily by females.

The next (unpublished) studies were the 1972 thesis works of Jane L. Dorn on ravens in Wyoming and Roderick N. Brown's 1974 writings on the Alaskan raven. Brown described over thirty call categories, including the "nestling kaah," "juvenile kaah," "antagonistic kaaa," "kruk," "krrk," "nuk," and "kwulkulkul.' He gives spectrographs of each of the different sound categories. But, to my eye, there is sometimes more variation within the sound categories than between them. I recognize few of these calls and have difficulty matching them to either my sonographs of Maine ravens or others' sonograms.

More recently Richard N. Conner from Virginia Polytechnic Institute recorded eighteen call types of the raven in southwestern Virginia: "*Caw*, growllike, whine, rattle, *cawlup*, staccato *caw*, *awk*, *cluck*, *kow*, bell-like, *ku-uk-kuk*, *ko-pick*, *awk-up*, *woo-oo-woo*, uvular, *o-ot*, *puddle*, and ke-aw."

The calls I eventually published in sonograms, because they were important to describing specific behavior of the birds at baits in Maine, were the "quork," "yell," "trill," and "knocking." Given the written descriptions of raven sounds and their sonograms, my quork seems to be Gwinner's "kra," Conner's "caw," and possibly Brown's "alert kaww." My "yell" is probably Gwinner's "rüh," Brown's "juvenile kaah," and Dorn's "ky," but I am unable to identify a comparable sound in Conner's work. My "trill" is also unidentifiable in any others' work. Finally, my "knocking" is Gwinner's presumed white-stork imitation, probably Brown's "kwulkulkul," and Conner's "rattle" (although his sonogram shows a rattling frequency of 30 hertz, or cycles per second, while mine is half that). In 1945 Francis Zirrer described the distinctive call as "a loud metallic, bell-like note, well imitated by the striking of a light hammer on a heavy piece of tin." The mechanical rattling or knocking calls are common in the females of many species of corvids in sexual or self-assertive situations, as Derek Goodwin has described in *Crows of the World.*

The "yell" sounds to me very much like the call the juveniles make when they want to be fed. Roderick Brown heard these calls given by birds with their head feathers fluffed out ("possibly to conserve heat") while looking around actively. He thought they served as a stimulus for feeding by the adults and signified the subordinate status of the caller, thereby decreasing the potential for aggression. (I saw them given by dominant juveniles, and my interpretations differ radically from Brown's.)

The quorks, which Gwinner considers a mild warning call

given when the birds are threatened, seem to be what the Australian corvid specialist Ian Rowley considers a type of territorial advertisement call. Roderick N. Brown observed it in displaying dominant males and thinks it functions to alert others, because the others often left "for no apparent reason" when one of the dominant birds gave it. Michel Andrieux has provided a sonogram of a raven's "warning call," but I am unable to identify it with the sounds I have heard or seen published. Tony Angell also describes "sentry" ravens uttering an alert "kaa." But in five years of observations I saw no evidence that ravens post "sentries."

There are many nuances of the quork, and I suspect that there are individual differences whereby resident birds recognize their neighbors. Alternately, there may be specific calls that signify as-yet-unknown meanings with regard to territory.

Vocal mimicry is another source of call variation. Gwinner's separate groups of captive ravens developed group-specific dialects. In addition, individual-specific calls (often mimicry of other sounds, such as a dog's bark or a grouse's call) served the ravens as "names." For example, if a raven's mate was removed, the distraught partner called for it by mimicking the missing mate's individually specific call. A similar sound-object association was shown by Gwinner's raven Wotan, who had repeatedly been called by *"Komm"* when fed tidbits. *Komm* means "come," but to the raven it probably meant "food," and the bird later called *"komm"* to his mate to bring her close to feed her.

The ability of ravens to "talk," to mimic human words, is possibly related to calling the name of a loved one. The birds learn to mimic the voices of those close to them. Singly reared ravens who have contact only with their human keeper and not with other ravens readily learn to imitate that person's voice. In the wild, ravens rarely mimic other species of birds.

Other, seemingly innate calls include a soft "korr" used at the nest, which immediately causes still sightless and apparently sleeping young to gape. An oft-repeated "rrack" near the nest causes the young to duck into the nest mold. The call is obviously a warning note, one that is audibly quite distinct from other calls.

Birds are primarily emotional beings, and their responses to emotional drives are probably much more direct than ours are, since human reactions are tempered by reason. Emotions are more "primitive" than reason, and I presume that many animals have very similar emotions to ours but, since they are driven by emotions, to a stronger degree. One major expression of emotions is vocalization, but there is no *a priori* reason to suppose that when a raven feels sad, it makes a sound that sounds "sad" to our ears. It could just as well sound "happy."

Many animals make arbitrary sounds that, like codes, have specific meaning. Thus, the mating calls of different grasshoppers, cicadas, or birds are very distinct, and to our ears they have no emotional content. Similarly, other calls of a sparrow, dove, or warbler also have little meaning to us except through the intellect when we figure them out. It surprises me, therefore, that many of the raven's calls sometimes display emotions that I, as a mammal for whom they are not intended, can feel.

When a raven pair is intimate with each other, they make cooing noises that *sound* soft and tender. When a situation arises where I expect a raven to be angry, it gives deep rasping calls that convey anger to my ears. I also feel I can detect a raven's surprise, happiness, bravado, and self-aggrandizement from its voice and body language. I cannot identify such a range of emotions in a sparrow or in a hawk.

Both Konrad Lorenz and Tony Angell describe the invitation to join in flight given by their pet ravens Roah and Macaw, respectively. According to Lorenz, to invite a family member to follow, each raven flew close over its keeper from the back at good speed, and in passing it wobbled with its tail and called a sonorous yet sharply metallic "krackrackrack." (Roah also tried to lure Lorenz away from places the bird avoided, but instead of the rack-rack-rack calls, it mimicked its own name!) Nothing is known about the context of this behavior in wild ravens. However, Nelson quotes Koyukon native hunters: "If a raven sees people hunting, it will occasionally help them find game. It flies ahead, then toward an animal that is visible from above, calling *ggaagga—ggaagga* (animal—animal). He does that so he'll get his share from what the hunter leaves behind."

And "hunters also see a raven tuck its wing and roll over in the sky to show where bear or moose is standing." Is this folklore or fact? A trapper/hunter from Nenana, Alaska, with whom I talked at length about ravens volunteered that he had seen the display. (I had not mentioned it to him.) He told me that a raven, after dipping right and left, fanning its tail, dived over a place in the woods while making a specific call. After the raven repeated the maneuver, the man had "the distinct impression the raven was trying to show me something." He followed and discovered a moose! Coincidence? Do ravens guide wolves and other "dangerous" animals to hunt for them? It is possible, but I am personally skeptical. I later suspected this trapper had read Lorenz more closely than Nature, and when I asked him months later, in a different context, I found out: "Yes, I have read *all* of Lorenz!" I dearly wished that he had not; as a scientist, I cannot trust complex observations that are fitted into a pre-existing mold of the mind. The weight of consistently repeatable observations themselves have to create the mold or the idea.*

* *As this was going to press, I received a letter from Paul Sherman, a well-known behavioral ecologist from Cornell University, who told me about an observation that he had just made in Idaho (while studying ground squirrels). Sherman is a trained observer who has no expectations of raven behavior. Sherman writes:*

My study site is a huge cattle ranch, and two nights ago a heifer died. The rancher put the carcass out in a pasture for the coyotes. The next morning I was walking out to my study site, and I stopped to look at the carcass through my binoculars. It was about ¾ mile from me, over rolling, open ground.

When I focused the binoculars, I noticed one raven about 2 m. from the carcass. As I watched, it silently took off and flew directly toward me. I put down my binoculars and watched. The raven flew up over my head and when it was right above me it gave several throaty "croaks." The bird circled me twice, then slowly began gliding back toward the carcass. I waited and presently the raven returned overhead and repeated the circling and croaking performance. Then again it sailed off toward the carcass, this time going all the

So far, no critical studies to support or reject the suppositions of these very interesting anecdotes are available. Lorenz,
citing no evidence, maintains the display is meant for other
ravens to follow. If the trapper and the Koyukon hunters are
correct, it is meant for dangerous predators instead. Can the
bird use the same display in two totally different contexts for
entirely different purposes? Whether or not ravens lead hunters, follow them, or both, and what they say and how is a
fascinating mystery that remains totally uninvestigated.

Ravens also sing, as do many other corvids. As Arthur
Cleveland Bent writes, "More than any other bird I know,
the raven will converse with himself for hours at a time, a
curious gargling, strongly inflected talk." Francis Zirrer calls
the song "a soft musical warble" and "a most remarkable
jumble and medley of very pleasing sounds." He often heard
birds in the fall singing in unison while perched on the tall-
est trees, a soft, drawn-out musical warble containing "lisps,
croaks, buzzing sounds, and gulps" in a refrain such as "spor-
spree-spruck-spor-per-rick-rur-ruck." The song is heard usually from the middle of August through late fall.

We have hardly begun to decipher the language of the raven. Its dictionary so far contains but a few "words." Perhaps
our analysis has been too coarse-grained to catch the meanings. Our research has been something like that of aliens from
outer space who make sonograms of human vocalizations under different situations—eating, playing, loving, fighting, etc.
Certain differences noted in frequency, intonation, and loud-
ness are correlated with feelings and emotions. But human
sounds convey much more, and perhaps ravens' do, too.

---

*way back and landing. After it landed, the bird looked back
several times at me (I was too far away to hear whether or
not it vocalized again).*

*I had the distinct impression that this bird was alerting
me to the presence of a carcass. It also seemed to me that it
was, in essence, showing me the way to go. There were no
other ravens in sight and I could see a long way in all directions, this being a huge open pasture.*

THE RESIDENTS KEEP IT ALL

NOVEMBER 2, 1987. The leaves are all down, and we have already had two snowfalls of several inches each. Now, however, the weather is mild. There is no wind, and it drizzles at night. The moisture brings out the nutty scent from the newly fallen leaves that have just turned brown. Luminescent green patches of moss decorate the rocks and the ground under the firs and spruces. Some of the wild apple trees have a few yellowish fruits hanging like isolated decorations. But down in the remnants of the old orchard below the camp, there are few apples left. The bears have picked them. To do that, they climb up into the trees, and since they can't reach to the ends of the limbs, they simply bend them and break them. These broken limbs on the well-pruned trees still have dead leaves. All the others have dropped.

It is the beginning of the fourth winter's raven season. I make a big circuit up to Gammon Ridge, to get reacquainted. This year not a single red oak or beech in the area bore nuts. (The next year the beeches had one of the biggest crops I can ever remember. The blue jays were caching the nuts constantly, but I saw no crows or ravens feed on them.) Usually the beech trees show fresh bear-claw marks, their branches are broken like the apple trees', and the ground below is all pawed up by bear and deer digging for nuts. Not this year, *anywhere*. There are no squirrels up on the ridges, either. But down in the valleys, where the balsam firs grow, there seems to be a population explosion of red squirrels. I've never seen so many. They are harvesting fir cones, churring from all directions, and here and there you see them chas-

ing each other. One pair almost ran over my feet. The chickadees also seem to be superabundant. I come upon one flock
after another of a dozen or more each, accompanied by
downy woodpeckers, pairs of red-breasted nuthatches, golden-
crowned kinglets, and occasionally brown creepers. I heard
a pileated woodpecker. There are big flocks of finches—red-
polls, pine siskins, goldfinches, and evening grosbeaks—in the
hardwoods.

I first hear and then see two ravens flying together over
Center Hill. One makes the usual deep quorks. The other
makes shorter, more rapid ones. Suddenly they circle down
toward the trees. Out pops a third raven. It has been silent,
possibly hiding. The two renew the vigor of their quorks and
shoot down the hill after the third one. I can follow the chase
with my eyes for about half a mile. One raven is the most
active, trying to get close, the second of the pair seems content to tag along behind. I have often seen groups of three
ravens flying "together." Had I not been here to see the beginning, and not had the background of the past three years
of fieldwork, I would have seen a "group" of three ravens.
Again a matter of interpretation versus observation.

NOVEMBER 3, 1987. At 7:58 A.M. a single raven comes leisurely flying by. It suddenly back-flaps, makes a slight turn,
as if checking on something to make sure, then flies on, silent
as before, as if not wanting to let on it has seen anything. I
know it has seen that thirty-pound pile of meat I put at the
edge of the woods by the apple tree some hundred yards from
camp.

The second meat pile, farther down the trail and out of
my sight, is also discovered today, but here the response is
quite different. I am alerted by almost continuous, very loud
quorking. And there is *only* quorking—no yells, no trills.
Usually the residents are silent at their meat. But today I
hear the undulating quorks of a bird belonging to the Hills
Pond pair and the deeper, very rasping calls that belong to a
bird from the Grahams' Farm pair. The meat pile is between
the two nests, but closer to the Hills Pond area. When I go
down to check up on the commotion at 3:30 P.M., only the
Hills Pond pair is directly at the bait, and several pounds of

meat have already been removed. Occasionally I also hear the knocking sounds that I now know are given, primarily if not exclusively, by females. I have heard the quorks in late summer every morning from near the nest area for the past four years. I now believe the quork or a variation of it (there are many nuances) is the territorial advertisement call. What I am witnessing here is probably a contest, where each pair proclaims to the other that this food is in *its* territory.

NOVEMBER 4. Near dawn a raven that remains silent comes by the meat pile at the camp. All is also still silent near the pile below where the two pairs were yesterday, and I go there quickly to hide in the thick pine woods on the hill above. At 6:30 A.M. I hear the Hills Pond bird coming from the left, from the direction of Hills Pond, and within a few seconds after that comes the deep-voiced one from the right, from the direction of the Grahams' Farm. I do not see them, because they do not get as far as the meat. Soon their calls are fainter as a chase ensues. Then, from nearby in the woods, I hear a few "chunks." Did one stay? The call sounds like an abbreviated version of the female knocking call, or maybe the call I had previously interpreted to be the "all clear" signal when I had left nest areas.

At 7:10 I see a raven being aggressive toward another that is sitting on a tree branch. New quorking. There are still no yells, no trills. The quorking, the acoustical territorial marker, is still being broadcast at 8:20.

No meat is eaten. At 3:00 that afternoon I suddenly hear the Grahams' Farm bird near the bait. Through the thick branches I barely make out another raven plummeting down toward the bait from near Mount Bald, giving the high-pitched quorks. It is soon quiet, but when I check the bait before leaving, I find no birds there.

NOVEMBER 5. THURSDAY. Excitement today! Four ravens arrive at dawn, and one of them is making quick and continuous quorks (*not* the long, loud territorial advertisement calls). They all circle over the meat and settle in the woods nearby. I soon hear the juvenile yells. But within two minutes I also hear the first territorial quorks as the Hills Pond

pair comes in. The birds quickly disappear from my view in
the woods, but I hear an agitation call, signifying that one
of the adults is aggressively challenging one of the four newcomers.

At 6:25 A.M. the four show themselves again, flying over
the bait very closely and perching in the trees nearby. One—
and only one, as usual—is giving the juvenile yells. I can see
the bird's pink mouth. The Hills Pond pair is perched nearby
in the same tree they perched on last winter. Now, strangely,
the newcomers fly over to them, as if *attracted* to them! Do
they perch near them to check out the enemy's resolve? Do
they want to know if these are friends? In any case, the answers are given: Twice one of the pair lunges at the birds
that come close to them. The victim gives the agitation calls
and flies off.

At 6:45 the newcomers leave, one by one. But ten minutes
later the youngsters are back for another quick fly-by. The
Hills Ponders again give their distinctive territorial quorks,
and then all leave for the rest of the day. I know these four
are not young belonging to this pair, because it, like most of
the others, experienced nest failure this spring.

The Hills Ponders fly down to the *other* meat pile. Now I
hear lots of quorking and knocking. Then all is silent. I stalk
down through the woods and find old W 4 and her mate feeding (W 1, the old mate, has since long been replaced by a
new male). No other birds are in the vicinity. So they secured the bait for themselves after all. The Grahams' Farm
pair has been driven off this lower bait, and four juveniles
have been held off the upper one.

Thinking I can force them to take a greater interest in the
meat pile up here, where the juveniles are still trying to feed,
I bring up all the meat. Forty minutes later I hear the Hills
Ponders' typical quorks down below where the bait had been.
They are back. There is also an accompaniment of the short,
quick quorks (the female?). And a *third* call—a very choppy
series of fairly continuous quick quorks that I have not heard
before. I presume it has something to do with surprise or
agitation because the meat is gone.

Those four birds coming in this morning seemed unusually
eager to feed, but the resident pair was there almost instantly.

I predict the four will be back next dawn, this time with reinforcements! And *then* they'll feed.

NOVEMBER 6. I awake before dawn and look out—to see snow! The ground is covered, and it is still snowing. Lighting a match in the dark, I see the temperature is 20°F outside and 50°F inside. My new stove and the additional oakum chinked between the logs of the cabin last summer have helped considerably. I make my fire, have my coffee, and go out to sweep the snow off the meat.

At six o'clock dark clouds come drifting by in the gray dawn. I'm eagerly watching at the window. 6:10 A.M.—They should be here now. Any minute. I'm excitedly waiting to see the crowd. Nothing. 6:14—Ah! W 4 comes up from Hills Pond! She is silent and perches by herself. She perches on the same aspen the pair habitually used last winter. After five minutes she makes a few of the high gonglike quorks. A second raven flies by, silently. It cruises fairly high and gives no hint that it has seen anything. It is not W 4's mate, because it has a conspicuous split tail. I have not seen it before, at least not recently. Split-tail could not have missed seeing either the meat or the attending resident bird.

At about 6:32 W 4 gives a few of her typical knocking sounds. Split-tail comes by, and W 4 takes off to give chase. I hear the agitation calls in the distance as they disappear from sight. But Split-tail is back again for still another pass at 6:44, silent as before.

For two hours I see or hear no ravens, but at last I hear juvenile yells in the woods. Did Split-tail hide in the forest? Immediately afterwards, I again hear the agitation cries, rolling, rasping angry quorks, and then the Hills Ponders, W 4 and her new mate, come to feed at the bait and carry off chunks of suet to cache. As I would now predict, they are totally quiet. Two hours later, the two are *still* caching and still alone. Most significantly, they are still silent. No other birds have been near.

Today, I heard none of the distinctive territorial quorks of the pair. The Grahams' Farm pair was repulsed yesterday. And today the single vagrant, Split-tail, didn't stop, or if it did, it was chased off. Even the four roving individuals that

were here yesterday as a group have moved on. The residents
have successfully defended their meat, and I am very fortunate to have seen the process.

NOVEMBER 26. Today is Thanksgiving Day, and I'm again
driving out to Maine. I never seem to miss a storm, and just
now I ran into the biggest snowstorm of the year so far. It
started in Maine last night, and when I get to the hill in late
afternoon, it is still snowing heavily. Ten inches of light
powdery snow have accumulated. Broad white cushions of
snow are bending down the tips of the branches of the firs
and spruces, so that the trees look like sharp pointed spires.
The dark green underneath the white seems a dull black, and
the sky is gray, punctuated by millions of white flecks slowly
drifting down. The nearby hills still show a dark gray, and
the distant ones are barely visible. During the long drive I
keep thinking of the big black shapes of a pair of ravens
flying low over the pointed spires of firs and spruce, but none
is to be seen. Will there be many to greet me on the hill?

As I walk up to the cabin, I hear only the crunching of the
snow under my feet and an occasional high "tsee-tsee" of a
ruby-crowned kinglet. There are no ravens. I scrape away
the new snow from where the meat pile was left on November 5 as part of the experiment. (This is uncharacteristic,
because except for specific experiments, I usually pick up all
the meat when I leave, not wanting to have a permanent
feeding station, which would be highly unnatural.) What I
find will tell me a lot about what went on in the intervening
three weeks: whether the pair has successfully defended its
meat, or whether the two have been swamped. What a surprise—the meat is still in place after twenty-one days! If I
had not watched with my own eyes when I first put the meat
out, I would have guessed (as I did years before) that no
other ravens had been in the area, but I had actually seen at
least seven others, and they had all been very much interested in this bait.

I am more and more attracted to my hypothesis. I cannot
provide an absolute proof (unless I could possibly capture
the pair at just *the* time when others happen to be in the
area—a tall task) because I am not dealing with an abstract

mathematical formula but rather with an intelligent, highly flexible animal. The raven's behavior at any one time is influenced by many variables, and I cannot be sure what they all are. The best I can do is to make observations again, again, and again, under the same conditions, and under as *many* conceivable other conditions as possible, until there is a consensus or an internal consistency or both in what I see. When you get many observations, some of them often seem totally incomprehensible and seemingly contradictory; but if you later find an underlying pattern that unites them all, you see beauty! The more observations the pattern pulls in, the greater the beauty. Ultimately I have the feeling that just *because* the pattern is beautiful, it is also true. I've been told that this is almost literally so in mathematics. But with these birds, I first have to be absolutely certain it is true before I can allow myself to feel the beauty.

NOVEMBER 27. I'm awakened to a rosy red dawn under a crystal clear sky with temperatures at 10°F. I rush out to uncover the meat.

At 6:38 A.M. a raven flies over, then a second one. The pair has come—the Hills Ponders. They quork a few times during their apparent morning inspection for intruders and return down the valley to the pond. For the next three hours I see only the ever-present blue jays. They have not made a sound all morning on their frequent trips to the pile of new bait.

At 9:45 I suddenly see several ravens. I cannot count them because in the next half hour they circle over only briefly, disappear behind the trees, return, circle some more, and disappear again into the forest. One flies to a tree where another has landed, and the first leaves; the second flies on to another perched bird, and that one leaves also. Two circle the bait together. I hear one set of deep quorks and one set of knocking sounds. There are no juvenile yells and no trills.

None of the birds who came at 9:45 was marked, and from the set of deep quorks I heard, I suspect that the Grahams' Farm pair is among them. But between 12:08 and 12:37 I see only the Hills Pond pair (W 1 is indeed gone or else has lost his tag). They fly over several times, always

silent, possibly because the other birds have left. At 12:37
the female (W 4) leaves by herself, making knocking sounds
in flight as she recedes into the distance. At 2:55 a single
unmarked bird circles over to investigate briefly, and after
that there are no ravens all day.

Throughout the afternoon the fire in my wood stove warms
the cabin, and swarms of cluster flies awake from their winter slumber and gather at the windows. I let volleys of them
out, and they fly across the clearing and fall down on the
snow or land on the trunks of the birch trees near the cabin.
The blue jays switch from feeding at the bait to catching
flies. A northern shrike comes and watches from the top of
a maple tree. I see it regurgitate a pellet, which I later retrieve (it looked indistinguishable from an owl pellet, mea-
suring 2.5 x 1.2 centimeters and contained bones and mouse
fur). I had previously seen a shrike pursue chickadees, and
all of the chickadees now leave, silently. The shrike chases
flies instead. Each time I let out a volley, it catches one or
two in the air.

I do not believe that the ravens did not return because
they saw me. But I will find out. I leave after dark to visit
Charlie and family in Jefferson, to spend the next day deer
hunting. Maybe the ravens will go down to feed then.

NOVEMBER 28. Five of us get up at 4:30 A.M. to go on the
last deer hunt of the year. It is also my first this year. After
strong coffee, toast, and scrambled eggs, we drive to where
the deer tracks are thick. We take up our positions after
walking for a mile in the dark and wait for the sun to rise
as we strain our eyes and ears and shiver. We are too charged
up to notice hunger, and we do not even think about food
until well after dark when we finally get home. As usual,
none of us gets a deer. Some of us heard a crash here and
there, and someone saw a flash of brown. Mostly we saw fox,
fisher, weasel, and coyote tracks, as well as enough deer
tracks to make us think one was hiding behind every bush.
But the most notable thing I saw was a raven. I heard it
first. It was flying very high, all by itself, making continuous
happy-sounding, gurgling-quorking noises. I have no idea

what they mean, but I call it "singing." The raven flew on, and I heard the happy sounds recede into the distance.

When I get back up the hill in the moonlight, I read the snow. Only blue jays and crows had been at the two baits. I know now that I was not the one who had kept the ravens away.

NOVEMBER 29. At 6:30 A.M. the eastern sky is ablaze just above the horizon. Flocks of finches dash like clouds of pepper across the sky, but I see no ravens. By 7:05 the sky colors have faded, and it is getting light. Two crows have flown in from the north and are sitting silently in the red maple above the bait, facing the rising sun. By 7:50 the two blue jays and the northern shrike are back. Still no ravens, and none shows up until 10:00 when one flies over, and I take leave from another raven-watching session.

DECEMBER 18. A foot or so of light powdery snow muffles my footsteps as I come up the hill with a burlap bag over my shoulder carrying a hundred pounds of meat. The air is crisp and cold, near 10°F. And it is very quiet. Not a breath of air.

As I come near the birches after the first steep rise, I find a tuft of feathers on the snow. Finch. A *tuft?* Plucked! There must be more. Making a detour in the woods, I find them; they have been blown about by the wind, so it did not happen today. Under an isolated fir tree I find the heavier wing and tail feathers—pine siskin. So this is where the bird was plucked, in the cover of the evergreen branches. Could a raven have caught this bird? No, a raven would not have gone to this dense fir tree to pluck and feed. It was a small predator that likes evergreen thickets, probably a saw-whet owl because the sharp-shinned hawks do not stay here in the winter. The finch was probably caught in its sleep.

I continue up the trail, into the stand of large fir trees and spruces. Here I hear the high "tsee-tsee" of a brown creeper. The bird looks like a brown mouse, running up a tree trunk in jerky little hops.

There are no ravens, or signs of any. Coyote tracks mean-

der over and around where one of the baits had been. I set
out new bait for tomorrow.

DECEMBER 19. It is so quiet this morning it is almost eerie.
Not a single bird sound. Not the tiniest rustle of a breeze.
The only things I hear are the deer mice scampering about
the cabin in the predawn, their most active time. When I
hear them, I always know it's time to get up. They are more
reliable than my alarm clock.

There is no sign of a sunrise. During the night the stars
were veiled, and now there is not even a trace of the sun.
The whole sky is a uniform gray, and I see the first tiny
flakes of snow slowly drifting earthward.

8:30 A.M. A single raven comes from the east, gives several quick, deep quorks (not the longer territorial quorks),
sets its wings in a dive, and then flies on as if it had seen
nothing! It is snowing heavily now.

9:10. I play the raven's territorial quorks over my loudspeaker. Nothing comes. A good excuse for me to walk in
the snowstorm. Snowstorms are hard to resist in the forest.
You can feel the essence of the North Woods.

I walk north. In the hardwoods going up to the ridge the
ground is peppered with the hulls of white ash seeds. Looking up, I see a flock of eighteen pine grosbeaks busily feeding.
I did not hear them, but when I stay totally quiet and watch
them I can just barely make out their almost constant low
whispering. After nine minutes the birds suddenly take
flight, making very loud, constant calls, "chee-chip" or "chee
chip-chip." There are no more seeds on the tree, but about
five hundred seed hulls are left on the fresh snow. One wine-
red male, etched against the gray sky and the falling snow,
seems to have been left behind. He calls loudly, turning his
head in all directions. The flock calls far off in the distance,
and the bird flies directly toward them. Does the calling of
these birds in flight serve to keep the flock together? Is
this why geese continually call in flight when they are
migrating?

There are many white ash trees with seeds in these woods,
and I soon spot another flock of pine grosbeaks. They, too,
are quiet except for the low whispering call. I time this

group for ten minutes; not one loud call. At sixteen minutes the flock is joined by eight newcomers, which give loud calls while they are in flight and as they settle and join the others in the food tree. As soon as they settle, they are also quiet. They erupt to make loud calls only when they finally leave as a flock of fifteen.

I notice other flocks. A dozen or so evening grosbeaks fly over, flaunting their brilliant yellows and flashing black-and-white wings. While in flight they constantly make their clear bell-like calls. The biggest flocks are those of the pine siskins and red polls. These flocks—fifty to a hundred birds per flock is common—also chitter in whispers when they feed, but as soon as they take flight, they make two calls. Colors are too muted for me to distinguish the siskins from the red polls. But their calls—"trr—trr" and "cheet, cheet"—are fairly loud.

As I come within sight of the cabin, a crow perched high in a maple tree caws three or four times, and I see a second crow fly up instantly from the bait. There are no other birds, and the second crow could not have seen me approach.

12:15 P.M. By a lucky chance, I'm here to see a *second* pair of crows discover the meat pile. As I expected on such an occasion—which I have not seen before—there is lots of cawing. Only these caws are different, because the birds do not leave. Soon the cawing stops, and both pairs are caching meat. At one time all four are briefly down at the meat together, but for the most part they act as if they are mildly annoyed with each other. After about an hour both pairs appear to have adjusted to an amicable existence. There is no more cawing. Finally they totally ignore each other. As they take breaks from caching and feeding, a silent blue jay jumps down to feed. Two hairy woodpeckers occasionally loiter in the trees near the bait, but they are too timid to go down and feed with the jays and crows, always waiting until they leave.

1:48, 2:25, 2:28, 2:30: A single raven comes four more times, flying furtively close over the trees, making just enough of an appearance to see the bait and then departing. Each time the crows caw, and after the raven's third appearance, they all leave. None come back for the rest of the day. At

2:30 it is already beginning to get dark. In two days it will be the shortest day of the year.

DECEMBER 20. Another gray dawn, like yesterday. I predict it will be another snowy day. I also predict that there will be several ravens here today, because the single bird that came to check out the bait yesterday was almost certainly not a resident bird. It was probably the same bird because it quorked only once, the first time, when it flew by at high altitude. The other visits were always low over the trees and always silent. The Hills Pond resident birds never showed up. If they had, they would have come from the opposite direction. The loner should bring recruits this morning.

At 7:30 A.M., I see four ravens in the vicinity! None of them is marked! And as I had expected, they are not at all quiet. They make many short quorks in rapid succession. I also hear a number of "thunks." There are no long territorial quorks or yells.

7:38. The newcomers have given themselves away in their enthusiasm! I hear the Hills Ponders' quorks—the long, loud territorial advertisement calls. Both birds come up from Hills Pond. Sorry, recruits, no feeding now. Four is not enough of a crowd to get at this lunch.

The Hills Pond pair stays around until at least 8:45. At first they call frequently and fly conspicuously all around the area before perching in the top of the red maple; their ears erect, side by side, they begin billing each other. Others are perched lower, in the thick branches of the same tree. One of them vibrates its tail in a submissive gesture, pulls its neck in at the nape, opens its beak, and makes appeasement sounds. One of the pair flies down to it, and the submissive one leaves. I suppose the submissive one has just pleaded, but to no avail.

An unmarked bird flies up to the bait at 8:43 and 8:51. This one does *not* do the typical jumping-jack procedures. Given my work with the captive birds, I know that it is experienced, probably an adult. It is down again at 9:08, and a second bird tries to join in, but a scuffle ensues and the second bird immediately leaves. At 9:20 and 10:13 the unmarked bird is down again but does little or no feeding.

The Hills Pond female flies over and joins him; he hardly acknowledges her presence. This means that he is her mate, just as I had suspected. The pair does not feed; it looks as if they are just keeping tabs on the bait.

In the afternoon it snows hard, almost a white-out. The Hills Ponders leave. Suddenly one bird goes down and does the jumping-jack maneuvers. This is therefore a cautious, inexperienced juvenile. It is also a bird that is unfamiliar with this strange bait. It has not fed here before. Others join it, and three ravens are feeding silently in the storm. None of them is marked.

DECEMBER 21. Today the Hills Pond pair comes separately. The male, arriving at 7:00 A.M., remains perched high in a maple tree near the bait. He has the erect ears of a dominant territorial bird as he swivels his head in all directions. He is not just waiting, he is actively patrolling. And he is not patrolling for anything on the ground. This is very clearly guarding behavior, not bait shyness—I am now more confident of my interpretations.

At 7:41 a group of three unmarked ravens flies over. They are silent and quickly fly beyond my sight. The Hills Pond female arrives at 7:59, and after that I hear singsong vocalizations mixed with knocking sounds from the woods. They continue until 9:30. Between 9:47 and 10:34 I see eight separate fly-bys by single birds, and on four different occasions I also hear yells. All of the fly-bys are brief, silent, and low, almost through the trees. These birds appear furtive. During all this time the resident male perches in the aspen and scans the area continuously, as he did earlier in the morning. At 10:21 I hear a defensive call from the woods. Maybe one of the pair has taken direct action against an intruder.

At 10:55 the Hills Pond pair flies together over the bait. They fly high and are very conspicuous, and I hear their distinct territorial advertisement call before they leave for Hills Pond.

At 11:20 a raven lands in the top of a spruce tree beyond the bait, puffs out its throat and elevates its ears. It must be one of the pair. It sits for a few minutes before flying down the valley to Hills Pond and coming right back with a white chunk of fat in its bill! It eats it in the birch close to the bait, then flies back down the valley to Hills Pond. There are no more signs of ravens for the rest of the day. No wonder the territorial birds can afford the time to guard all day without going down to feed: They are living off the meat they have cached here previously.

Reviewing the results, I read the morning's events this way: The Hills Pond pair stayed to guard the bait until 10:55 after three intruders, who checked on the bait at frequent intervals, had finally given up and left. The male came back for one last check at 11:20 and, finding no birds here, left almost immediately.

It is still hard for me to accept the fact that not one raven fed here today simply because of interference. I counted twenty-one fly-bys. Clearly they were very much interested in the meat. Is there still a tiny possibility that they are scared of the meat or me in the cabin? I do not think so, but the remote possibility makes me nervous.

DECEMBER 22. Today's results were almost identical with those of yesterday, and the nagging question I had is answered. Yesterday I threw some stale hot dog rolls out the front door. They were lying against the front step. No raven fed at the meat pile some hundred yards from the cabin, but one raven *did* surreptitiously walk along the side of the cabin and remove the hot dog rolls from near the front door! A bird walked in the snow next to the *back* door as well, where there was no food.

It has not been a bad week. I have learned that although three birds are not enough to force an entry to a resident pair's bonanza, they may do so by stealth, such as under cover of a blizzard. I suspect there are probably not enough ravens around right now to recruit. Possibly they are at a moose kill in Rangeley or elsewhere. Or is this a small social group that has decided it has a chance to make it on its own without recruiting competitors?

WHY BE BRAVE?

Bravery means taking risks for a higher cause (i.e., for the social welfare or at least for some delayed reward). I am trying to give at least a crude definition of bravery, because I want to discuss whether or not some ravens are brave, and why. No bravery is possible without fear. And we have seen ample evidence throughout this book that ravens, especially juveniles, are extremely fearful. That is, they avoid risks. Ravens are never thought to be foolish, and indeed it is in no small measure because they *avoid* risks that they are seen as intelligent.

If we can assign fear to a raven without labeling it an "anthropomorphism" (and I will risk it), we can also bring up the issue of bravery versus foolishness. Throughout this book we have seen that ravens both take risks and are social or at least gregarious animals. My tame ravens showed an almost paranoid fear of any fuzzy or feathery object and of *any* "new" or strange object; some of the birds *always* avoided these and other new and risky things until another individual had thoroughly checked them out. The interesting part is that while one bird always avoided them, another went eagerly *out of its way* to check them out, even after I had purposely satiated it with the tastiest food until it could not swallow another bite, to be sure it was not taking the risk because of hunger.

The vagrants in the field face risks. Dangers can lurk at any unattended carcass, and the first raven at it faces the brunt of the aggressions of the defenders. Clearly this suggests that there is a great value secondary to feeding for a brave raven. But what is that value to the individual? Most immediately, of course, it benefits others to have brave individuals near. In our society, too, we value bravery because it ultimately brings benefits to us, and we have figured out a way to encourage it. We pin special medals on the brave so they can be readily identified and thus receive paybacks in higher status and its attendant benefits. It would be surprising if bravery could long endure in ravens if brave individuals were not in some way identified and rewarded. I examine the possibility here in terms of both the risks and rewards.

In addition to the danger of approaching a still body that may or may not be dead, a second, even more immediate danger involves the carnivores already feeding on the kill. How much should the birds risk to feed from a wolf's kill? In general, carnivores are reported to be remarkably tolerant of ravens. Richard Nelson writes: "Wolves seem to pay little attention to ravens at a kill. If they [ravens] find wolves at a kill, they land almost amongst them, flap around the meat and foul it with excrement. Sled dogs act the same way, as if ravens hopping around them did not exist. Even the wolverine is said to do nothing to drive ravens off that land beside it and steal its food."

Perhaps that is statistically the most common scenario. But that kind of statistic is not relevant to a raven, to whom it is enough to be caught and killed only once. The 99.5 percent of the time that the wolf is tolerant hardly matters. In Maine, ravens are generally not injured by coyotes, but they can be. Dave Lidstone, a logger working near my study site, once found a spot where a coyote had killed a raven on the snow close to a moose carcass.

Despite evidence to the contrary, a raven is likely to risk danger by approaching meat of any kind, not just meat that may not yet be dead. By far the majority of carcasses that ravens have contact with in the wild are provided by carnivores, and these large mammals—bears, wolves, coyotes, and foxes—are usually still in the vicinity. Generalities aside, they sometimes guard their meat even against ravens. My studies have shown that when a carcass *is* recruited to by ravens, they can eat most of the meat in a very short time.

Unless the carnivore who killed the prey also guards the carcass, it will soon have to hunt again. It follows that one can expect active meat-defense by the carnivore. Indeed, Laurel Duquette, who did her master's thesis on the "Porcupine" caribou herd in Alaska, observed wolves "jumping up to snatch at the ravens" who were trying to share the kill. Jane L. Dorn saw ravens give way to coyotes at carcasses; if a raven sidled up to snatch a bite, it was lunged at. Pat Balkenberg of the Alaska Department of Fish and Game told me he found grizzly bears protecting carcasses by sleeping *on* them, presumably to protect them from ravens. The bears swiped at and sometimes charged the ravens, generally "to no avail, because the ravens are too agile." Ellen Hawkins describes an episode when an injured wolf in Minnesota sought to defend two deer carcasses: "The ravens are getting braver, and he [the wolf] can hardly stand seeing them at a deer. They come dropping down out of the trees around whichever deer he's not at, and he has to hurry over to get them up. They scatter briefly, but here they come again, settling down around the other deer, and back he has to go." As the wolf became weaker, it conceded one carcass to the ravens and defended the other by lying on it.

Moreover, wildlife ecologist R. O. Pedersen saw a wolf kill a raven at a moose carcass. And, as I have mentioned, one of the wild ravens that I marked in Maine is now in a roadside zoo, because it got injured after being caught by a wolf in an open enclosure. Clearly there is a risk in getting too close to the carnivores at a bait, or else ravens (at least in Maine) would feed at carcasses at the *same* time as coyotes, instead of alternately.

Now comes the mystery. *Despite* the risks, *some* ravens seem to go out of their way to court danger. The Heinroths observed that their pair of tame ravens were both extremely skittish and extremely bold. For example, a flag hung three hundred feet from their cage sent them into a frenzy, and for hours they banged into a windowpane to try to get away. However, they attacked spectators and also the large animals at a zoo, including elk and bison. The Heinroths' explanation was that the birds were "testing" potential prey to see if it might be weak and could be eaten. I reject this hypothesis for at least three reasons: (1) The ravens were well fed and therefore not likely to be motivated by food; (2) the birds were not likely to know that a bison was good to eat; and (3) the behavior emerged when they were adult and not before.

Ravens must approach the bait to get food, and it might be argued that the bravest are the hungriest. Often, however, this is clearly not the case. Wildlife photographers Jim and Kathy Bricker, who were photographing wolves in Ontario at carcasses set out near a blind, saw interactions of ravens with a bald eagle and a fox. Ravens repeatedly sneaked up behind the fox, who always "shooed them away." And this scenario took place no matter how long the ravens had been feeding before the predator arrived. Even after the ravens were done feeding and the bait was unoccupied, if another predator came, the ravens *repeated* their approaches to it. The Brickers observed that "the two birds that were the bravest were also those who had *not* been feeding before the predators arrived." Kathy Bricker writes: "I honestly get the impression that challenging the intruders was not so much from hunger as from desire to impress one's fellow ravens with one's daring. Perhaps these are males doing their chivalrous bit for the females." I have examined the Brickers' filmstrips of the ravens interacting with a bald eagle at a bait, and there is no doubt that the brave birds are strutting and feather posturing, doing the classical raven dominance display of courting males.

David Bruggers of the Bell Museum at the University of Minnesota saw six ravens land fifteen feet from a bait in northern Minnesota where a rough-legged hawk was feeding: "The largest of the ravens strode toward the carcass and the hawk, then cautiously sidled closer, striking the carcass feet-first near the hawk, and then he (presumably a male) displayed his 'ears' and ruffled his shaggy throat feathers, and strode slowly back to his companions, who now gathered screaming around him in a semi-circle, while he began making choking sounds while bowing and raising his wings. Eventually the smaller ravens stopped their screaming, and he then walked back to the carcass and repeated the entire performance." The "choking movements" are the typical bowing movements of the male courting raven.

A more direct support for the boldness-to-enhance-status hypothesis comes from another carrion-feeding corvid, the European carrion crow, *Corvus corone*. A study by Tore Slagsvold of the Zoology Museum in Oslo, Norway, showed that when he placed a stuffed eagle owl on the ground near where there were crows, there was male-male *competition* for the privilege of attacking it! Usually a single crow acted as attack leader and tried to chase the other harassers (technically "mobbers") away. It looked as if the crows were competing for the privilege of showing off. Slagsvold placed stuffed male and female crow dummies near the owl, and the most active of the crows attacked not the owl but the male crow dummy. In ten out of eleven cases where he succeeded in killing the most active mobber, it proved to be an adult male. These results have been difficult to accept because they are at odds with all mobbing theory. Nor has there been any theory so far to indicate that males enhance their status and mating success by showing "bravery."

"Bravery" in corvid birds is very well known, but it is explained in the literature under "play" behavior. For example, wolf researcher L. David Mech specifically cites the raven's "playful" behavior:

> As the pack [of wolves] travelled across a harbor [on *Isle Royale, Michigan*], a few wolves lingered to rest, and four or five accompanying ravens began to pester them. The birds would dive at a wolf's head or tail, and the wolf would duck and then leap at them. Sometimes the ravens chased the wolves, flying just above their heads, and once, a raven waddled up to a resting wolf, pecked its tail, and jumped aside as the wolf snapped at it. When the wolf retaliated by stalking the raven, the bird allowed it within a foot before arising. Then it landed a few feet beyond the wolf and repeated the prank.

I have seen my pet crows act similarly with a dog, which was not amused. It would be very interesting to find out if corvids act this way without an audience nearby, and

whether or not acts of daring result in elevation or maintenance of status or mating success. I do not question that the
ravens could have been playing. Indeed, I think "play" is
the obvious, immediate "explanation." But there is likely
an ultimate evolutionary significance to such daring behavior, which may directly or indirectly translate into mating
success.

Consider the case of the raven in the winter: During this
season when ravens breed, the male must provide carcasses,
at least for part if not all of the food supply during the incubation period. What would determine his acquisition of
food? First, the ability to hold a territory, which should depend on *dominance* over other birds who might contest
either the territory directly or the food bonanzas within it.
Second, the ability to *find* carcasses, which in turn would
depend on good eyesight and vigorous flight to cover large
distances. Third, daring (or experience) to approach objects
that could be food without being killed in the process. All
of these qualities would be essential in a mate, if one hoped
to reproduce, and it would seem that there is one simultaneous proof of all these qualities: If a male can bring a potential suitor *to* a carcass, and show her that he (and hence *she*
as well) can feed at it. But carcasses are rare; there is not
always opportunity for such a demonstration. So the male
shows his mettle on substitutes. The bird who approaches a
live wolf will surely dare to approach a carcass. The ultimate
proof of his worthiness is the food he provides, and in ravens,
eating often depends on bravery, and bravery is gained
through experience. In raven society it separates the "men"
from the "boys."

TRADEOFFS AND COMPLEXITIES

*Back into the chamber turning, all my soul
within me burning,
Soon again I heard a tapping somewhat
louder than before.
"Surely," said I, "surely that is something
at my window lattice;
Let me see, then, what the threat is, and
this mystery explore—
Let my heart be still a moment and this
mystery explore. . . .*

—EDGAR ALLAN POE,
The Raven

JANUARY 9, 1988. Before going up to camp I stop at the Grahams to ask, "What's new in Franklin County?" Mike chuckles and says, "Nothing. Unless you count a cow that died in my barn."

I go immediately to the woods in back of his house to find the brown Guernsey, now frozen rock solid. She is totally untouched even though she is within a quarter mile of the Grahams' Farm raven nest. After an hour with an ax I have plenty of exposed meat. Now the ravens can begin their feast.

If ravens come, I want to see whether marked ones from last year show up and how the resident pair interacts with the crowd. I need to find out more about which birds cache meat and which vocalize. I want to take photographs. Lastly,

I want to watch for any of the unexpected things that almost always show up and give you a different perspective. There has never yet been a day I've watched ravens when I didn't see something new. This is what keeps me going. As Pascal said, "Perfect clarity would profit the intellect but damage the will." I will have the will for many years to come.

JANUARY 14. Unlike last time, there are no ravens at the Bethel dump as I go past on Route 2 and I am very curious to see what I will find at the Grahams' cow.

We have had a record cold snap. Temperatures are still near −25°F and the deep snow is very "dry." It does not wet my legs, so I do not mind walking in it, especially when I am about to see a crowd of ravens. Sure enough, about ten fly up without sounding any alarm when I come close. The cow is about half eaten.

Brent Ybarrondo and Jim Marden, my two graduate students from Vermont, have come with me to help install a proper wood stove for heating (about time!) and to lug up 500 pounds of meat. The cold motivates us to get the stove set up in a hurry. It *works!* We get the upstairs of the cabin to a toasty 60°F. This is surely a record. The only problem is that the new stove has been coated with some kind of black paint that gives off noxious fumes. So tonight for a few hours we are outside in the snow, under crisp, starry skies, roasting our steak on the grill over an open fire at −20°F. We've brought "Johnny" in a flat little bottle, and we pass him around. "He" is soon in out of the cold. And that steak was the best of the many we have had here.

JANUARY 15. The stove is still warm in the morning. Unprecedented. I marvel at how survival was possible the last three winters without it!

At 9:10 A.M. a raven flies silently in to inspect the 500-pound meat pile and perches in a tree for a few minutes before it flies on. Three blue jays, two hairy woodpeckers, and one white weasel with a black-tipped tail take their turns feeding on the meat.

Later in the morning we check on the twelve baits that were laid out over thirty-three miles at the same locations as last year. The three farthest baits have been totally cleaned up by very large crowds, and judging from the tracks it must have been at least two days ago. Two more are just beginning to be fed on by crowds, with about half a dozen birds flying up as we get near. Six have been visited by only one or two birds. Coyotes have fed at one bait that had the large crowds. A pattern of results almost identical to last year.

JANUARY 21. Leaving Vermont at 5:30 P.M., I drive all night in the dark and get to the cabin by about 10. Climbing up, I have a hunch that the 500 pounds of meat will not have attracted a crowd. If it had, I would have found tracks on the trail. I am right; there is not a single raven track at the meat. What a disappointment. Strange how I want to see ravens, when a lack of them is surely just as informative. It indicates that gatherings are not just random accumulations.

JANUARY 22. I awake at 7:00 A.M., and a few minutes later, surprisingly, I hear ravens. They call several times, and I see eight fly by. After making an inspection, they disappear. Clearly they knew about this bait. A minute or so after the fly-by I see two ravens chase another. Then all is silent.

I go to the Grahams' cow carcass, which I had buried under snow to keep it. Still, ravens have dug down to part of it and are feeding at that spot, but only two fly up as I come near. I work for over two hours digging out the rest of the carcass and building a blind some ten yards from it. The blind is made of twelve small fir trees stuck in the snow, and on the inside I weave a network of fir branches between the trees. In front of the peephole I have made an eight-foot-long alley of fir trees, so that my camera lens will be hidden in the dark. In two days I'll be ready to get a picture of ravens and to see them up close!

In the late afternoon, after nothing comes all day, I go to see Larry Wattles, who will take me to the top of Taylor Hill where Anne Moody lives. Mrs. Moody has reported to Peter Cross, a Maine game biologist who wrote to me, that she has seen red- and white-marked ravens. I told Cross about my marked ravens, and he has been relaying sightings back

to me. These are especially significant because they have led
me to a roost—and, as it turns out, a relatively "permanent"
roost.

Larry Wattles, the Moodys, and Clarence Nutting, a trapper who also lives on the hill, all confirm that the roost has
been there for at least two years every fall for one or two
months. The roost is fifteen air miles, as the raven flies, from
my cabin, and given that the birds swerve around Mount
Blue, which is directly in between, it is in the direction consistent with many morning arrivals last spring.

I'm also told that the ravens are not there every night.
Tonight is one of the nights they are away. Wattles tells
me that when they settle in, just before dark, they make the
most amazing noises he has ever heard: "It sounds like a
circus." All is quiet tonight. I'm ecstatic. I've learned something else. I've been wrong about my idea that all the roosts
here are temporary. The temporary roosts are likely to be
near temporary food bonanzas far away from the main roost.
(This was later corroborated many times.) From this roost
the birds have access to these dumps.

JANUARY 23. It is a wind-still night with a quarter moon and
a mass of stars. A coyote's wail-moan-bark comes from Gammon Ridge. I answer in my best imitation. The wolf-coyote
answers. And so we call back and forth.

I'm up early to see if any ravens come by to check out the
meat again. But I'm not hopeful because I suspect the birds
who came by yesterday morning went right on to the Grahams' cow, which is probably their first choice. And indeed
no ravens come by as I watch until 9:00 A.M., although one
calls.

The cow draws the activity. As I get there at 9:30, about
ten ravens fly from the area, and there are signs of feeding.
I crawl into my cold and cramped hiding place, where my
only view is through a 400-millimeter lens.

After about two hours, I hear heavy wingbeats. A raven
lands nearby in a tree and starts up an amazingly varied
vocal display. But it leaves after only ten minutes.

I am getting colder. My right leg has fallen asleep, but
just as I try to stretch, I hear wingbeats again. A raven has

landed very near. Then another, and another—I think six
to ten have arrived. They are fairly quiet, except that I hear
a soft gurgle here, a pure-toned tinkling liquid call there,
and then suddenly the loud juvenile yell. Another bird joins
in. Now they are quiet again, and I hear pecking all around
as the birds are doodling on the branches. A twig falls, a
piece of bark. Ravens, unlike crows, are incessant "doo-
dlers"—they keep their bills constantly busy when they are
idling. Then, suddenly, for no discernible reason, they all
fly off.

A half hour later a crowd gathers again. One bird goes
down finally. Sleek and alert, it walks to the bait in the flat-
headed feather posture. Another bird comes down almost
immediately. This one is a fuzzy-head, so it is not its mate.
The sleek head attacks the fuzzy head. The fuzzy-head jumps
aside and takes off as if in sudden fright. The flat-head takes
off, too.

Four more times the group gathers, and each time I hear
yells but no trills. Twice only two birds come, and when
there are only these two, I hear both knocking and trills but
no yells. Later there are two birds (at least one with a pink
mouth) with one trilling and making the demonstrative male
posture in front of a female (bowing motions and knocking). A pair of juveniles courting!

When groups are present, I usually hear territorial calls,
probably challenges from the resident pair. At numerous
times there are also vigorous chases and the defensive calls
ravens give when they interact agonistically while perched
in the trees.

Twice more a single flat-headed bird goes down and walks
near the bait, but none of the group follows. For the fourth
time a bird flies to the ground about fifteen yards from the
bait in the woods and starts to walk toward it through the
underbrush. I can hear its footsteps crunching on the snow.
But it has not walked more than two or three yards before
I see through a small peephole in the needles another bird
flying down and driving it off. Then all the birds leave, and
I again hear the chase calls in the distance.

After five hours, at 2:30 P.M., it is totally quiet again.
There are no ravens nearby, and my feet are numb. I can

get up and leave. It feels great to run through the woods in the heavy snow, and I return to the cabin exhausted, but I do not think it is from the run. The five hours of sitting still in the cold have taken a toll all their own.

I don't know why they did not come down to feed. I suspect two things are occurring rather than just one. If my work with the tame young ravens last year is any indication, it is the youngsters who are most afraid of the bait. Perhaps they are in a quandary: They would like to join the dominant adult that shows the bait is safe, but they may be attacked if they do. The adults may have had an easier time keeping the juveniles off today because the juveniles had already been spooked, or because they were still suspicious of me or my lens, or because they had already fed at one of the other baits.

Given the balance of risks and rewards, the vagrants and territory-holders who seem to be in conflict may in the long run actually help each other. The highly dominant territory-holders can feed at any bait the vagrants find that they themselves might have missed; they are never excluded. Indeed, under cover of a crowd they can even feed on their *neighbor's* bonanzas, which they would not know about except through the crowd's activity, and to which they would not otherwise have access alone. Also, if times get tough, the resident pairs may abandon their territory, and then they have the option of joining the vagrants at the roost.

As I see it, the vagrant strategy is the basic one, the bread-and-butter raven operation. If you are a territory-holder who can make it on your own, you are a rich "aristocrat"; you can be overrun at any time, and perhaps will be, but there is normally inhibition against it. And all of those who overrun you aspire to the same position themselves. Theirs is not a different strategy. It is just a contingent strategy: Do *A* if possible, and if not, go on to *B*. If these complexities are possible, then "inconsistent" observations should not be unexpected, especially in a highly evolved system, or in a bird possibly intelligent enough to weigh the consequences.

JANUARY 24. Perhaps they didn't stay to feed yesterday because they saw me enter the blind. Today I'll beat them to it. I get up at 5:00 A.M., have breakfast, and walk into the woods at 6:00, carrying two hot bricks wrapped in cloth, sandwiches, and a piece of log to sit on. I try not to run so I won't get sweaty because the 0°F temperature will be a killer if I sit motionless all day in damp clothes.

I make it inside the blind with my luxuries by 6:35, and it isn't much too soon. At 6:50 I hear the heavy wingbeats. It is still dark and thus safe to pull the branches aside; I see six ravens perched in the trees around me. There may be even more. They came silently, and they remain so. All you hear is the birds' pecking the branches, shaking their feathers, and drawing their beaks through their feathers to preen. It sounds papery, like turning the pages of a book.

At 7:09 there are a few short, rasping quorks. Several birds now leave. A minute later the rest follow, silently. They are heading east, and very quickly I hear chase calls in that direction. Is this *it?* How come there were no yells? Why didn't they show even *signs* of wanting to get close to the bait?

7:18. A raven appears alone, gives a couple of rasping quorks, and within five minutes hops down to feed. This must be one of the resident pair. It feeds as if it hadn't a care in the world. After seven minutes I hear what sound like the territorial advertisement calls. A second bird flies down to join the other without any hesitation. The first doesn't miss a peck, and the two feed as if oblivious of each other. Definitely a pair. Both have the "smooth" feather posture of undemonstrative birds.

They continue feeding silently until 7:36. It looks to me as if they waited until the juvenile crowd left. Maybe they guarded before that. I would love to see what would happen if the juveniles came back now!

7:41. My wish is coming true. Two more birds arrive, then a third, then six are down feeding. One or two are yelling. More birds fly in, but these others—twenty, maybe thirty—only sit in the trees. Then three of them yell almost continuously from 7:44 to 7:59. I count about one yell per second. That makes at least several hundred yells. The rest still don't go down. One raven on the very top in the center of the cow assumes its intimidating stance—"pants," ear tufts, and flashing nictitating membranes—but I'm not sure if it is a new individual or one of the pair that has now changed its behavior. An instant later they all fly up in a clatter of wings. And all the rest in the trees take off with them. What happened? Did one of them finally see the camera lens?

I think ravens are extremely sensitive to anything round and shiny, such as an eye or a camera lens. Eyes have special significance. Gwinner describes how, although ravens fight viciously for dominance, they never peck one another's eyes. During squabbles I see them routinely feign pecks to the eyes, but they never follow through. Yet, when they find a large dead animal, they always peck out the eyes. With my captive ravens, eye contact was extremely important for any interaction. When I had been away for a few days, they crowded close to the window to see me, but as soon as my eyes wandered away from theirs, they found other things to do.

It is 8:00. All is silent. A red squirrel churrs nearby, and a woodpecker drums. No ravens for over two hours.

At 10:02 I suddenly hear a *very* vociferous raven in the distance coming nearer. It perches only briefly, giving high-pitched continuous rasping quorks strung into each other in an unending succession, varying only in emphasis. Interspersed are occasional gurgling-knocking sounds. Within a minute it flies down to feed. I see the lining of its mouth numerous times through my camera lens (I do not snap the shutter today because I am afraid of scaring the bird away), and it is *definitely* a bright pink—a juvenile. However, its deep, rasping voice is worthy of any adult. Is it trying to set up a territory? After finishing its feed, it hops up into the crown of a tree above the bait and continues its loud vocal display. At 10:12 it leaves. For the next six minutes I can still hear it calling while it flies and then perches some half mile away. Eventually the calls fade into the distance.

At 10:45 it is back, making only a brief stop before again going off on its rounds. Surely the residents could not possibly have missed this bird, which made no attempt to be furtive. Why is it tolerated? Is it one of their offspring from a previous year? What is it calling for?

At 1:50 P.M. two ravens fly in, silently. I hear very soft gurgling-knocking sounds. After a minute the birds leave, as silently as they came. The resident pair, checking?

2:10. One bird comes silently, perches in the trees for fifteen seconds, and leaves silently.

2:46. I can again hear the familiar noisy one coming from afar! It behaves just as it did the two other times, but this time it does not come down to feed.

At 2:56 it comes back still another time, perches, calls some more, and keeps looking around. What is it looking for? Three minutes later it leaves, continuing to call in the distance, and it keeps this up until 3:06 when it comes back again. You can track it by the noise. Meanwhile, six ravens have come, silently. I start to hear the yells, and the two yellers have pink mouths. Where are the adults now? Weren't they attracted by all the commotion? The six go down to the bait and feed from 3:06 to 3:27. None had a fluffy head or the adult strutting behavior. I saw no agonistic interactions. There were no chases.

After feeding they perch in the trees, and at 3:43 they leave silently, one by one. It is starting to snow, and dusk is coming on. Just as I'm about to crawl out of my hut, I hear a funny noise in the tree above—a beak-snap quickly followed by a gurgling hiccup, and these sounds are repeated a dozen or so times. Then the bird makes several rasping quorks and leaves in the direction of the others. It is 4:01. I have no explanations for what happened today.

Why were there no more adult interactions here at the bait? Are the adults reluctant to leave their immediate nest territory because it is now more important to hang onto *that* than to defend a specific meat pile? I had heard the deep rasping territorial-advertisement calls of the Grahams' Farm pair at their nest site numerous times throughout the day. Once these calls were almost continuous for half an hour, and they made them near their stand of pines even when the vagrant crowd was up here near the cow. When alone, these birds are usually quite silent near the nest.

I also heard calls in the distance that sounded like the Hills Pond pair, making me doubt that I had been observing individual variation of voice but rather a very specific adult call, or do neighboring birds mimic each other? If so, why?

But what can possibly explain the bizarre difference among the juvenile vagrants? Why was one that was all alone and showed no fear of going down to feed by itself so blatantly noisy, and why did it come back again and again without feeding? I am disturbed because there are so many complexities that I have not yet begun to fathom. But even if I find all the "rules" of their optimum behavior, that does not mean that every individual will heed them. After all, do humans?

JANUARY 25. This morning I expect to see them take the bait. I can hardly wait to get out to the blind again. It has snowed overnight, and I brush the cow off.

By 6:34 A.M. I'm relaxing inside the blind. The birds come between 6:45 and 6:48. They come silently except for one who gives a few long rasping quorks when it gets here. After that, until 7:05, all you hear is the papery sounds of them drawing their feathers through their bills, and lots of feather shaking. One would think there were bird baths in all directions. It seems they are again here simply for their morning toilette. If they are interested in feeding, I should hear lots of yells, but I heard only one bird yell very briefly. Are the resident pair guarding?

Suddenly a raven makes a few low, soft grunting calls, then another does, and another—and they all take off toward the east as they did yesterday. I think the short, soft, low grunts are a "follow-me" or flight signal, because I have also heard it as they approach a bait. Are they off to another bait?

All is quiet for thirty-five minutes, then two very noisy birds come. They make agitated quorks, fly around making a quick inspection, and leave fifteen seconds later, returning to the Graham's Farm nest area. There I then hear a few trills and lots of the stereotyped knocking sounds.

At 7:58 the by now familiar "singer" can be heard coming from afar, giving the same concert of rapid, strung-together quorks, gurglings, and knocking sounds. It keeps this up for fourteen minutes and leaves for the west, whence it had come. Today it doesn't even feed here.

I wait until 9:00. Nothing more comes.

The results are not what I had expected. I did not expect the noisy juvenile would *continue* to be alone. And I did not expect so many birds appearing together. Are there individual vagrants as well as small bands of them?

JANUARY 28. It snowed about a foot two days ago, and all the fresh signs will be visible. Yes, they have been feeding heavily at the cow for the last three days. And the meat pile by the cabin looks the same. Tomorrow I'll finally get my pictures.

Just to be safe, I get up at 5:00 A.M. again to be in my blind by daylight. Unfortunately I forgot how hard it is to move in deep snow, and it turns into a race with the ravens as I try to run in the snow in the fast approaching dawn. When I make it to the blind, I am drenched in sweat, and I collapse exhausted. But I'm on time. I don't hear the sounds of raven wings for another ten minutes. As before, they perch in the trees and preen. Then one of them makes long rasping quorks. Not good. The bird sounds agitated. Sure enough, in another minute they all leave. (Because of me or the quorker?) It is not even 7:00.

In the next two hours the cold becomes almost unbearable. I do my best to do isometric exercises, but I am soon near the end of my tolerance. The crowd did not come back; only one or two totally silent birds and a pair making the knocking and the territorial quorks came. I am mystified as to why they don't feed. Well, there is always the bait in the field at the cabin.

As I had hoped, there are about twenty ravens at the cabin when I get there. One sees me coming up the trail and makes short, nasal, gruntlike sounds. I pretend to walk *away* from the cabin as they all leave, before I return and enter. By 10:30 one has approached the meat twice (I watch through a slit in the rug over the window of the cabin that has been darkened out). After the second attempt, not yet having touched the meat, it perches in a tree and sings.

At 11:35 I hear the deep, rasping, territorial quorks and the more sing-song Hills Pond quorks, and I see a chase with a white-marked bird in pursuit. The white-marked bird heads on down the valley, and within twenty seconds a pink-mouthed juvenile is feeding! Four more now come close, but the feeder stays only half a minute, looks up, and flees in apparent alarm. All the others leave instantly as well.

I have two hundred pounds of meat in reserve down in the jeep. Maybe now is the time to make use of it in a little experiment. If the ravens are afraid of the camera lens, they should feed at a bait pile farther away. Tonight I make a second meat pile twice as far away from the cabin as the first. To carry the stuff through the deep snow up a steep hill for a half mile is tiring. After the last load, I'm drained, drenched in sweat, and panting heavily as I collapse into the snow wondering how many ethologists could duplicate these feats of weight lifting in the snow. Dropping into bed after working to exhaustion must be one of the most intense pleasures of life, because there is nothing at all that I would rather have done.

JANUARY 30. Thanks to my exhaustion yesterday I do not wake until 7:00 A.M. I will stay in the cabin today, anyway.

A group of perhaps eight ravens is close by at 8:30, and there are many juvenile yells. No feeding. I also hear the territorial quorks. The white-marked Hills Pond bird is here, giving an unmarked bird a vigorous chase. One bird stays after all the others leave, perching above the new bait pile. There it sings for about ten minutes. It has small erect "ears" and is unmarked.

Starting at 9:45 and continuing until 12:30, an almost flat-topped bird with small ears and a totally black mouth is very near the cabin and the baits, making the typical, liquid knocking sounds again and again. An adult female. I am surprised, because W 4 is the resident female. Later in the morning I see another adult sidle up to her while she is making the knocking sounds, and the two are very tolerant if not affectionate with each another. Are there then *two* pairs near this bait? Why do they tolerate each other as well as the juveniles? Do they fear the bait so much they would rather not be first down to test it?

After one set of knocking near 10:00, she flies down to the closer bait and flies right up again without feeding. At 12:30

P.M. I see at least four ravens nearby, and these four (with the "knocker" among them) make numerous passes over the area and perch very close to the bait. After half an hour they all leave. Three more make a brief appearance at about 3:00, and that is it for the day. I *have* to stay tomorrow, because all of this month's work to get photographs will be for nothing if I give up now.

JANUARY 31. I expect they will come early this morning, or not at all, so I am up early. It is the latter, except, of course, for the pair, but even here there is a surprise.

At 6:40 A.M. a raven flies over and goes to a high vantage point on the hill. Another joins it fourteen minutes later. I see them together twice and hear the knocking once. It is a pair.

At 7:41 the stillness is broken by the typical singsong territorial quorks of what I thought was one of the Hills Pond pair, but I see the *same* individual change its call to the deep rasping quorks of what I thought were the Grahams' Farm pair. Obviously I cannot identify individuals, but maybe they really do mimic each other. Right after calling, one bird flies down the valley, to be followed by a second, then a third. Immediately afterward there are chase calls.

Until 11:30 all is quiet, except that one or two birds come by briefly to give a series of the singsong territorial quorks. Now a pair is perched close together in the red maple. At first one of them does numerous series of knocking (I can see the lower bill go up and down with each knocking sound), then it preens the other. I also hear many of the soft endearment sounds. But, to my great surprise, I notice again that neither of the pair is marked. I had always thought this was the territory of W 4 and her mate. They had fed here exclusively before and vigorously defended their meat. Where are they now? Are they at a better meat pile elsewhere? Have they given up?

At 12:06 P.M. one of the present pair flies off silently. The other makes deep, rasping, agitated quorks, trills about ten times, and gives soft, short, nasal quorks. At 12:17 it makes very nasal honking calls and flies down the valley.

These birds showed not the slightest interest in the meat. Did they come up here to socialize? Clearly there will be no crowds here today.

Again no pictures.

THE CAGE RAISINGS

ANY OF MY QUESTIONS will probably not be answered by studying these shy, far-ranging, free-flying birds only in the wild. Eventually I will have to study them also in an enclosure large enough to allow them to express the behavior we hope to study, yet small enough so that we can see it and control some of the variables.

When you are on a hot research project, you can't leave it alone for months while you wait to see whether or not you'll get funding. Foundations hate something risky (i.e., interesting). If others haven't done something similar, it is often considered uninteresting or undoable. So I can't wait for funding to build the facilities. I do not have even the $6,000 to pay for the chicken wire for the giant enclosure. But I borrow it from the University, promising to pay it back when I get the money. The land, the hundreds of trees for the structure, and hopefully all of the labor will be free.

To build the raven cage, I planned a number of huge raven-cage-raising parties on the hill by Camp Believe It. I sent out invitations to the locals and to people as far away as California. Each party was a big event, with a roasted lamb as the main bait.

The first of these memorable events was held on August 31, 1987, and my first job was to get a lamb. Second, we needed a bulldozer to heave up to the building site the 133 75-foot-long rolls of 4-foot-wide chicken wire, lumber, nails, a window, a stovepipe, tarpaper, and other supplies. No lumber companies or hardware stores anywhere around had more than thirty rolls of chicken wire in stock. I talked to someone at a lumber company in East Wilton, Maine. Would they order the wire from the manufacturer and call me "immediately" if there were any problems? "Yes, we'll order it now." "Okay. I want 133 rolls." Two weeks later, the magic date is drawing near, and I still haven't heard anything. I call the lumber company again. "Has the wire come?" "No. We need to have your check before we can put through the order."

It wasn't easy to extract a check from the University that quickly; I had to get a bill from the lumber company first. Would they send me the bill from Maine by overnight mail? "Yes." Four days later the bill arrives, by regular mail. And so it went, back and forth, until Friday, just before the Great Raven-Cage-Raising Party, when I finally had the promise that my order would be ready to load at 9:00 the next morning. Danny Proctor, my 'dozer operator at $30 an hour, and I went down to load up. The order was not ready, and it was not assembled until 11:30—all that was there, that is. They did not have it all. Indeed, they didn't even have the miles of stitching wire that we needed to sew the cage together with. I learned that I have little patience.

We eventually managed to find stitching wire in stock elsewhere. Enough to get started.

Weeks earlier, I had cleared a road through the woods to the site, and now Danny and Dalton Proctor maneuvered their way up to the building site, dragging the trailer laden with lumber and ninety rolls of chicken wire.

Shortly after they left, it started to cloud up and then to rain. The forecast for the Great Weekend was more rain. It was the only rain for the past month.

Meanwhile, the skinned lamb was hanging from the big birch tree in front of the cabin. Swarms of yellow-jacket wasps were already tearing off hunks to feed their larvae. But the weather was cool. The meat would keep until tomorrow. I got busy with the chain saw, felling trees, and so we began clearing the terrain on the hill in back of the cabin. John Marzluff had flown in from Arizona, and Alice Calaprice from New Jersey. Both were soon well occupied in the rain, lugging logs out of the clearing. Other, more local friends lent not only present but past labor. Elsie Morse came by from the coast, bringing several, bags of delicious home-made chocolate-chip cookies, several fresh loaves of her special bread, and a scrumptuous mix of cereal for breakfast. The Swedish post-doctoral candidate Ola Jennersten, then at Brown University, came again, this time lugging a huge case of Moosehead beer. Another friend, Leona DiSotto, brought eight dozen doughnuts she had just made, and her husband Henry came with sleeves rolled up, ready to work. So did flocks of other neighbors and three carloads of students from the University of Vermont.

I was a little worried that a bear might come in the middle of the night and carry off the lamb, so footsteps outside at 10:00 P.M. gave me a start. It was Charlie, arriving early to be on hand at dawn. His friend Scott would be coming in a few hours.

I awoke at daybreak, hearing the quorks of the raven who *always* quorks his territorial signature. I listen for it every day. But this morning there was also another sound. Scott was stumbling through the door. He had never been up here before, and in the pitch blackness last night he made it only partway. When he started bumping into trees, he figured he had wandered off the trail. He finally felt a soft spot with his feet, and there he lay down to spend the night. His head happened to be right on top of an underground wasps' nest. With his ear close over them he at first mistook them for distant trucks lumbering along. He shifted position somewhat and slept, waking up nice and early before the wasps did and then continued on up. A true Mainer.

It was drizzling in the morning, but not enough so that I couldn't build a fire in the fireplace outside to boil the water for coffee. We all ate the crunchy cereal, fresh doughnuts, and cookies and eagerly went to work.

We soon had two crews working full blast. Ola supervised the crew building the observation hut in the center of the aviary complex, and John Marzluff from Arizona who would later come to work on the raven project with me, the cage construction. We worked steadily for two days. Yes, it *can* be done. Yes, it *will* be done. It is good to get started. I'm excited to think about the unique new approach to the problem that will be possible with these facilities.

It rains most of the day. But we are so into the work that
we hardly stop even for the heaviest squalls. Charlie keeps
the fire stoked in the big pit in the ground in front of the
cabin. At noon we put the lamb on a long pole over the coals.
Charlie's job now is to keep the fire going, to turn and baste
the lamb with our mixture of olive oil, lemon juice, oregano,
salt, and pepper.

In the evening the crowd, getting merrier by the minute,
gathers around the pit, and we dig into the lamb with knives;
we open cases of beer, and make short work of it all. The
skies begin to clear. The Milky Way shines like a brilliant
swath across the heavens as we move closer to the hot coals
to the accompaniment of Billy Adams' guitar. It has been
one of the best days any of us can remember.

By the third cage-raising party of the year we had most
of the aviary complex up, and John and his wife Colleen and
I finished the rest. People who see it now say: "This is *awesome!*" And so it is. It is a bird cage made from 133 rolls of
chicken wire that required cutting 225 maple trees just to
hold it up, and it is half a mile around the perimeter. It is
also, I know only too well, a big gamble any way you measure it. I cannot guarantee results, and I won't. But to me
research is an exploration and therefore always inherently
a gamble *if* it promises to uncover anything that is unknown.
So, for the time being, the project is only a gleam in the eye.
But that is not trivial—no sustained research of any kind
is possible without that gleam, no matter how sound the concepts or the plan. In the meantime, the field work goes on
as before.

THE LAST ROUNDUPS

> *"On this home by Horror haunted—tell me*
> *truly, I implore—*
> *Is there—is there balm in Gilead?—tell me—*
> *tell me, I implore!"*
> *Quoth the Raven, "Nevermore."*

> —EDGAR ALLAN POE,
> *The Raven*

THE PROJECT is now nearing the end of its fourth winter, and the mystery lies exposed enough to show a pattern, and many questions. If I waited until all the questions coming to light were solved before publishing, there would never be anything published, because the research would breed more questions *ad infinitum*. Somewhere one has to draw the line and sum up. This is a good time because I have gone much, much further than I had ever dreamed was possible with these elusive birds. The answer to the original question is at hand: Ravens recruit. In addition, I have some good evidence of how and why. These are discoveries worth reporting.

A detailed technical report is being sent off to Germany to Professor Hubert Markl, managing editor of *Behavioral Ecology and Sociobiology*. *BES* is probably the best journal to get the work read and evaluated by one's scientific peers.

Before it is accepted for publication, it has to be greatly shortened so it can fit into the very valuable journal space. It is hard to leave out so much and still retain coherence, much less a flavor of the ramifications.

I have concluded that there are numerous examples in nature where in the struggle for resources animals band together to overcome more dominant individuals holding the resource. But, of course, the ravens do much more—they actively *recruit* when they find a good resource. Their system may have things in common with certain human customs, which have evolved under similar ecological pressures. Take the !Kung bushmen of the Kalahari Desert. Each band of bushmen, living off wild plant food (*veldkos*, in Afrikaans) and large game, occupies a large territory. There are no social obligations to share the *veldkos*. However, the meat of the large game animals is distributed throughout the band. The band's bivouac is like the raven's roost from which hunting parties of one to five individuals go out in different directions, like ravens searching for large food bonanzas. Successful hunters recruit others to their kills, and thereby gain status and eligibility to marry.

Last year Marzluff and I prepared a research proposal for the National Science Foundation (NSF) to seek funding for a study of recruitment and the ravens' social interactions in the giant aviary that we are building. We want to test the hypothesis that recruitment confers high status, which is useful in procuring mates.

The bad news is that our grant didn't get funded. There were some excellent reviews and some detractors. Finally the panel said: "Few of the panel members were convinced that Heinrich could actually accomplish what he planned on doing."

The good news, however, was that I got supplemental funds from NSF so that I could follow up on the work already invested in the marked population of wild birds. I immediately called John, and he flew in from Arizona two days later. We decided to capture ravens and find out for sure if the birds behaved "normally" in the aviary. If so, we could make a good case that the aviary studies were feasible so that we could submit a revised proposal in time for the July 13 deadline, taking the reviewers' criticisms into account.

FEBRUARY 14, 1988. John has been staying at the cabin for a week now, and the bait I put out over two weeks ago has finally drawn a crowd of ravens. He has transferred the bait into the trap, and he called to say that the birds are going into it now. It is time for another raven roundup, so I come from Vermont immediately.

FEBRUARY 15. Near dawn we spring the trap and capture twenty ravens. As expected, the birds fly against the wire at first when we release them into the new big aviary, but within an hour some of them are already gathering to feed at the meat pile we have provided for them. Among them is Y 1, the sole surviving young bird from the lot I banded in the Grahams' Farm nest last spring. It is as aggressive and dominant inside the aviary as John said it was at the bait outside the cage a few days ago. Again one of our least vagrant juveniles is a very dominant bird. Might there be a pattern here to look out for?

In the evening the birds gave a performance that made me feel like applauding! All twenty gathered in a communal roost in the big balsam fir tree *inside* the aviary, making a tight cluster. A captive roost! We will be able to make an observation tower right next to the roost, to observe who perches with whom, and to see precisely how the birds that have had foraging success act and react. We will open a door (while we are hidden in the observation hut we have built adjacent to the cage) and let a raven into an adjoining aviary until it finds a hidden carcass there, either in a cage with a resident defending pair or alone. In the evening, when the raven scout wants to rejoin the roost, we will let it back in and look for the signals it may give. It will be like looking into an observation hive and watching the recruitment dances of bees!

MARCH 1. Great news! John has just called me telling me about his preceding twelve days' work at the cabin in Maine with the twenty ravens we captured. The trial run was a total success.

First, all of the wing tags held. It was good to have direct proof, to ease possible suspicions that the apparent disappearance of the vagrants from the study area was merely because of lost wing tags or leg bands. Best of all, the birds not only behaved naturally in our aviary, they seemed to like it or to feel secure there. When it came time to release them, John opened out a big section of wire and put the bait one yard outside the cage. The birds went out of the cage to feed. After finishing, they went right back inside, even though some of them had wandered around outside the cage.

The ravens' home life was apparently going normally. One mated pair preened together and stayed together. Two previously unmated adults who had had mates outside appeared to have started a new relationship. Clearly, ravens are adaptable. They had found plenty of food, and for them the cage was a good situation.

We gleaned one more bit of unexpected information about their social life. After the cage was opened, instead of all escaping, two ravens tried to come *in!* Indeed, they got in, so instead of twenty birds the cage now held twenty-two. The strangers were, however, treated with hostility and vigorously chased. That means that although ravens are vagrant, they quickly form (and apparently break) social ties to form and maintain groups. Here we had a demonstration of the same phenomenon as bait defense against strangers, but this time by juveniles that had been forced to be resident at a site which may have appeared to them "permanently" supplied with rich food.

I am very excited, because I am confident now that we will be able to remove doubt from the reviewers' minds when we resubmit our research proposal. The aviary studies are definitely going to be a tremendous boon to deciphering the details of the behavioral mechanisms.

MARCH 5. This may be the last raven roundup of the season, and as usual I need help. Also as usual, the students are eager, and some of the "hard-core" come out to lend their support.

When we get up to the hill in late afternoon, the place looks deserted. All of the meat that Henry DiSotto and John left is gone. The woods all around are trampled by the ravens' footprints, but there are no ravens. I feel we've come a day or two late. Will there be ravens tomorrow? We stay up late at night around the new wood stove.

MARCH 6. I'm in the blind at 5:30 A.M., ready for them just in case. A pair comes. One makes deep quorks, then switches over to the similar yet distinctly different, long singsong quorks. There is no yelling, even though there are about two hundred pounds of new meat in the trap. Many birds are gathering. Still no yells. Maybe the crowd is big enough, so no more recruitment is necessary. Every bird has access now.

Soon the ravens are going into the trap. At one time there are at least twenty in there, but as I get ready to pull the wire about five fly out. I wait until a few more fly in. Meanwhile more fly out. One bird makes calls that I have never heard before. I watch the traffic in and out for about an hour. I must have seen at least one hundred birds, but only one tagged bird—a green. That is a juvenile marked this year.

After about two hours of trying to get a larger haul of birds in the trap and watching for marked birds, I realize that the numbers in the trap are not increasing. I yank the wire, and the door slams down. The haul is twelve ravens. The age distribution is typical: eight juveniles, two subadults, and two adults. The mark-release task is now routine and uneventful.

Strangely, both of the adults we captured called when we released them. One (W 26), which flew in the direction of the Grahams' Farm nest, made the very characteristic high, short, staccato calls that ravens make when someone approaches the nest. Is it thinking of the nest to which it is now undoubtedly returning? The bill of this bird was short; it was most likely a female. The second adult (W 25) made long high calls. Both called repeatedly. This is very unusual; none of the eighty-nine others has ever given *any* call when we released them.

We finish with our marking and release all the birds well before noon. I have now taken the bait out of the trap and put it in a pile that we can watch from the cabin. We see ravens fly over the meat all afternoon long. Only one lands to investigate, then flies off.

I saw a long chase, the typical chase that you see near the *beginning* of a feeding cycle. This time I was very surprised to see that the *chased* bird was marked. Then I noticed the color: green, a juvenile.

MARCH 7. I presume that many more ravens will come this morning than came yesterday, because yesterday many of them fed undisturbed on a big food bonanza. As it is getting light, they start streaming in. I hear yelling. Before 6:00 A.M. about fifty of them (all unmarked) have gathered in the trees and on the ground around the trap where all of the meat was yesterday. I have now moved it up near the cabin where I can watch more easily. But these birds descend to where they had been fed rather than coming to the meat pile that they surely see or saw on the way in.

The meat pile lies untouched for at least an hour. It being early March, the crows are migrating. A group of about ten finds the meat pile, and within a minute they are down and start feeding. Seconds later a raven is down, then several more. The crows do not appear to be greatly agitated, but they hop to the side as the avalanche of ravens comes in. Soon there are twenty ravens, then close to fifty. The crows are by this time walking along the periphery of the crowd or sitting perched in the trees nearby. No ravens, as far as I can see, pay the slightest attention to them. They pay a lot of attention to one another, jumping and jostling at the bait.

There are almost no territorial quorks and no yelling. While they are all feeding, I hear one juvenile give the high-pitched calls that juveniles give when they are being attacked by superiors. This one bird keeps giving the complaint for at least fifteen minutes while it continuously wanders in and out and around the bait. It is G 2, a latecomer to the feast. As usual, there are the strutting cocks who can walk with impunity and feed anywhere they want. I also see birds with greatly fluffed-out heads display to these cocks by bowing down, flashing their white eye membranes, and doing the knocking calls. These are then single females courting the dominant (probably already married) males. At first I thought they were the males' own mates. But then I remembered that the resident females should be incubating now, and also that I never saw this female behavior when a known resident pair was present. In any case, the males appear to pay them no attention.

Earlier in the winter the real pairs looked entirely different: Males and females showed ears, and the females preened their males. Not now. These females do not erect their ears and stand tall; instead, they bow to the big males. My two young birds in captivity also bowed to each other, with the male displaying to the female in the "choking display," where the throat and head feathers were exaggeratedly fluffed, although there was no show of ears. Taken together, these observations reinforce my interpretation that the self-assertive display is *not* identical to the courting display as claimed in the literature. Back in the winter, the mated pairs did not display to each other; they both tried to intimidate the bystanders instead.

Not a single one of the other sixty-three juveniles we marked last year showed up. (But we saw four in the next winter.) Only *this* year's marked birds were at the bait this morning. In all there were only two of the five adults (W 8 and W 11) and three of the sixteen juveniles (G 2, G 12, and G 23) marked on February 17. (Another juvenile marked February 17, G 10, showed up the next fall.) Two of the juvenile feeders we marked only yesterday (G 27, G 28) were back today. All of this supports my previous conclusions that juveniles do not leave the area because they have been captured or marked. On the contrary, they are *most* likely to be seen back shortly after experiencing capture and marking at a bait. They drift away later, even though bait may continue to be provided. Rarely, some may drift back *in*, but that could be simply by random movement.

MARCH 11. I get up to the hill by about 4:00 P.M., well before dark. Only one raven flies up. Almost all of the 250 pounds of meat I had left four days ago is gone.

MARCH 12. I hear ravens at dawn. About twelve are down at the trap, picking on the remaining bones that have no meat whatsoever. The newly available meat is ignored for at least two hours. At about 8:00 A.M. one raven finally approaches the new meat, does jumping jacks, and flies up into a white birch facing the bait, where it sings lustily for about fifteen minutes. The bird is a juvenile (I can see its pink mouth), but it shows its "ears" and fluffs out its throat feathers just like an adult. Is it showing off?

Fifteen minutes after singing, it leaves and five ravens then come within minutes, circle over the new meat, then leave.

There is no further activity until almost 4:00 P.M. when an adult watches the bait from a tree nearby and makes knocking sounds on and off for nearly twenty minutes before descending to the bait alone. It is soon joined by another.

At 4:10 a juvenile perches on the same limb where the other juvenile sat and sang early in the morning. It looks at the bait long and intently and flies off toward the west, making a rapid series of calls in flight. It seems excited. I'll wager anything that I'll see large numbers of ravens tomorrow morning!

MARCH 13. I get up at 5:30 A.M. to put out a dead coyote, which the Wildlife Department at the University of Vermont gave me. I want to see whether or not the ravens will let sleeping (or dead) dogs lie. As it turned out, all the ravens but one (W 8, an adult) ignored it all day. W 8 walked up to it three times, looked it over closely, did the obligatory jumping jacks, but did not touch the animal.

As I had anticipated, at dawn ravens were perched all around in the trees. By 6:20 the first one had started to feed. It is G 37, a juvenile that we marked on March 6. More birds gather. In all, at least twenty feed here until 10:00. This time the initiator, G 37, is definitely not the most dominant bird.

There are only three other marked birds. W 8, the adult, could be unmated (or its mate is on a nest). This adult is indistinguishable in behavior from most of the juveniles;

G 23, one of the juveniles we marked a month ago, is always submissive; Y 1, the juvenile that was marked in the nest two springs ago, the sole survivor of six local clutches that year, is, as far as I can tell, the only resident juvenile. Unlike all the others so far, it is a "steady," having been sighted all around this area and up to fifteen miles away dozens of times by me and other observers. Does it remain resident because of its high dominant status? All juveniles, even this one, are submissive to resident pairs, but dominant juveniles can gain choice and possibly limited feeding spots among a crowd.

What I find very unusual this time is that the resident pair, W 4 and her mate, is again not present. They have not come the last two weekends. What has happened to them? Have they finally left here because the constant crowds are irritating them? (We saw them frequently the next winter.)

The next thing I find unusual is that for two days, with over twenty birds here much of the time, I have not once heard the juvenile assembly yells. Is this evidence against my recruitment hypothesis?

What it shows, I think, is how flexible these birds are. Having fed here for many weeks and always getting access to the food, they presumably know that they are assured of food here. They also probably know that there are many competitors for rapidly dwindling resources. Right now, only two to three pounds of meat remain. There is no need for assembly anymore. Besides, the defending territorial pair is absent. Maybe there is nobody to chase them away. Indeed, I have seen not one chase the past two days. Clearly the "constant" feeding this year and at the end of last year, with food left even after I've gone, has not been "natural." Ravens are otherwise rare. But that is the point; it has been another experiment.

OCTOBER 17, 1988. Today marks the beginning of the fifth season. We have already had one snowfall of over six inches here at Camp Believe It. It is foggy and drizzling lightly. The leaf-fall is at its height. The path up to the cabin is paved with crisp, freshly fallen yellow, brown, and red leaves. Tiny glistening drops of moisture cling to them. The path is more beautiful than any rose-petaled rug trod by a king.

I walk down the path, cross Alder Stream by stepping over the protruding rocks, and watch, from my blind, ravens coming to a bait. Seeing the thirty or so ravens interact, I again feel excited. I see it through new eyes. To me, the raven is now not just a bird, it is a *being*, masquerading as a crow, as it has been to almost all humans who have had extended contact with it. But I can now see a raven critically as well.

I can make predictions, and even the predictions are exciting. I predicted that none of the twelve young I marked in the nests this past spring would show up. And they didn't! I predicted W 4, the female resident we marked almost three years ago, would come. She did. I predicted that the birds would fly off in alarm if I walked up to them on all fours, dressed in a bear suit. They did. Now, if I could only impersonate a coyote!

As usual, one bird makes most of the recruitment yells. It erects its "ears" at intervals as it lunges at others who become "fuzzy-headed" in front of it. It is pink-mouthed. It is clearly one of the most dominant juveniles of this feeding group, if not *the* most dominant one. Yes—I "see" it all again, and much clearer now: It is the *dominant* unmated birds that yell; the unmated low-status birds (and mated adults) never do. Low-status juveniles probably recruit from a distance, but now in the presence of superiors, they have to let *them* take all the credit and eat at *their* discretion. It makes sense to me now. Dominants yell because they get to feed by attracting a crowd that overwhelms the much stronger resident adults. But once that has been accomplished they immediately fight for (and invariably win) choice of often limited feeding spots. The recruitment that results in sharing is *not* incompatible with subsequent fighting! And now I again recall from Gwinner that it is the dominants which preferentially get mates. I now feel excited, because I know I have a "program" to decipher the privileged spectacle before me. I am at an arena where dominance is established and held, and where ultimate reproductive decisions are made. Hundreds of seemingly disparate details have now merged into one simple pattern in my mind.

This day is a happy one for another reason. The paper I wrote on my work over the last four winters has just appeared in *Behavioral Ecology and Sociobiology*. All of that work reduced to fourteen pages of text!

Although I had read the manuscript many times, it had always been to find errors. Now I finally read it for pleasure. What I see is mostly between the lines—the cold dawns in driving snowstorms, the swaying in spruce tops, the spine-chilling climbs to nests on cliffs or in tall pine trees, the friends at raven roundups, the dead goat, the sounds and sights of the birds coming and going.

The research is going on and will continue to. A biological detective story differs from others in that the more you find out, the more you know you don't know. John and Colleen Marzluff and I have already begun the aviary studies. Caged birds require full-time attention, which I am unable to provide from Vermont. But I am jealous and want to hang on as long as possible. If I let others do the hands-on work, it is like somebody else going to the fair instead of me, and then telling me about how much fun it was. I don't like to live vicariously. But a sign of health in research is that it eventually leaves one's hands and assumes an independent life.

As I look back on the last five years, I'm happy that an ever-wider circle of people is getting involved, sharing the experience, and having fun. My efforts will be repaid if someone feels enriched by getting acquainted with an exciting fellow creature that is accessible to so many but known to so few. There will be more mysteries to solve— Evermore.

SUMMARY

THIS BOOK is about solving a riddle: Do common ravens, *Corvus corax*, actively disclose to strangers of their species the valuable and rare food bonanzas that one of them is lucky enough to find? If so, how do they do it, and why? I was highly motivated, because if the answer to the first question were yes, it would mean the discovery of a new biological phenomenon never before demonstrated in any animal. Furthermore, the "how" and "why" would necessarily be two additional, interesting questions that might be fun to try to answer.

In the body of the book I have described the *process* of exploring an exciting biological puzzle. Here I give the scientific results by providing numerous pieces of the puzzle, most of which were found during the course of the work. I also explain the logic of *how* these pieces fit together. No one piece is meant to be "final" and unalterable. Rather, the aim has been to discover enough sufficiently varied clues that are united by a consistent pattern.

The following eleven clues concern the main question:

1. At least 90 percent of the 135 baits totaling eight tons of meat through four winters in the woods of Maine and Vermont were consumed by crowds of ravens ranging from about fifteen to nearly three hundred birds.

2. The above numbers are noteworthy because ravens are not particularly common; on the average only one raven was seen per 167 miles of travel through raven habitat.

3. Crows, blue jays, woodpeckers, nuthatches, and chickadees also rapidly discovered, and fed from, the highly prized

meat piles, but none of the other birds gathered at the piles in large numbers. Therefore, ravens are doing something different.

4. According to the Audubon Christmas Bird Counts at four different areas around and within the Maine study site for the previous five years, only 118 ravens were sighted. The other birds were far more common, with both jays and crows being sighted at least thirty-three times more often than ravens, even though a raven can be seen and heard for far greater distances than the others. Therefore, the raven crowds in any one area or at a bait cannot be explained on the basis of relative abundance or random gatherings of passing birds.

5. Although the ravens passively displaced the other birds when they were feeding at a meat pile, many baits were for many months without ravens, and there was still no increase in the numbers of other birds at these baits. Therefore, the *lack* of crowds of other birds cannot be explained on the basis of competition from the more dominant ravens.

6. Blue jays sometimes gathered in groups of up to thirty in the spring, but there were never more than four at a bait. Therefore, although they have a mechanism of aggregating, they do not use it to share valuable food bonanzas.

7. Blue jays discovering a new bait vigorously defended it against others of their kind, so that no more than four birds remained. Therefore, defense of a valuable resource against others is a reasonable strategy in a corvid bird, and it *can* be executed successfully. Crowd formation is therefore not an inevitable evolutionary barrier that could not be circumvented.

8. Crows did *not* defend baits, but their numbers at the meat still did not increase, even though they were thirty-three times more common than ravens and roosted communally in large groups. This indicates that *passive* accumulation of birds is not likely to account for the raven crowds.

9. The numbers of ravens arriving through time were highly irregular. Typically, some baits had no crowds for weeks (and sometimes the whole winter), and then numbers increased to thirty or more in a day. These data are consistent with either chance arrival of flocks or recruitment;

they are inconsistent with individuals searching and arriving independently.

10. Realizing the far-ranging significance of distinguishing between the two hypotheses of 9, I spent over a thousand hours watching unattended baits until one or more ravens flew by and showed evidence of discovering the carcasses I had provided. In twenty-five instances of bait discovery, eighteen were made by single birds and seven by pairs. Crowds came at dawn only *after* a bait had been discovered a day or more previously. Therefore the hypothesis that the crowds are the result of discovery by flocks is rejected. (Although I do not exclude the idea that this can and does occasionally happen.)

11. Over four winters I tabulated the ravens seen flying over the countryside with no known baits nearby. Of eighty-seven birds seen, 69 percent were singles, 29 percent were in two's, and 2 percent were in groups of five to six. (Subsequently I also saw a group of over a hundred ravens in flight.) The percentage of the various groupings of ravens seen flying at large near the study area is almost identical to the groupings discovering baits. Therefore, the birds flying at large are also probably the same birds that discover the baits; that is, ravens forage by flying as singles or as pairs.

Given clues 1–11 above, I conclude that there is something "special" about the raven gatherings at carcasses (or meat piles, their artificial counterpart). The logical conclusion that can be drawn from 1, 4, 8, 9, 10, and 11 is that ravens actively *recruit*. The next question is: How do they do it? Here are a few more pieces of the puzzle for this part of the picture.

12. Groups of ravens are very noisy at baits. Indeed, I could easily locate any carcass in the forest simply by listening to and watching ravens. This suggests that mere commotion could be a signal that means "food here" to other ravens, because the alternative (to remain silent) is an option that is not used, although it is an evolutionary strategy shown in blue jays, which do not recruit.

13. A way to enhance the "commotion equals food here" connection through evolutionary time is to make the vocalizations more specific. Indeed, ravens at baits make a specific vocalization, the "yell," heard almost nowhere else. This call therefore very probably means "food here" to other ravens. (The yell most likely evolved from the begging calls that the fledglings give when they see an adult with food but cannot fly to it. It is as if the birds beg because they want to be fed since they see food, and others then "understand" why they beg. The *motivation* for giving the call need not be recruitment. It could simply be frustration when seeing food and not being able to have access, as when it is feared, guarded by aggressive individuals, or sequestered behind a screen.)

14. I recorded the juvenile yell, and one to five ravens arrived in one minute in eighteen out of twenty-two playbacks in the field, in the absence of baits but in areas where ravens had fed and were likely to expect to feed again (and where they presumably did not know of my concealed presence). No ravens had been seen for at least fifteen minutes (and often at least an hour) before the trial. Therefore, the birds were recruited by the call and their appearances were not random fly-bys.

15. Only one pair of ravens flew by in eighteen trials near baits of another raven vocalization, the quorks. Therefore, the birds distinguished between the two calls, and the juvenile yells acted specifically as the recruitment signal, while the quorks did not.

16. The attracting power of the yells was observed between different years at the same site with different groups of birds. Furthermore, in single trials the same response was observed from ravens in northeastern Maine and western Vermont, using recordings from birds in western Maine. Therefore, the call is not individually specific; that is, it attracts strangers.

The experiments (14, 15, 16) prove that the juvenile yell, which is normally given at baits, attracts other, strange ravens. However, as shown by the next experiment, it acts only in short-range recruitment (probably within less than a mile).

17. No ravens arrived in the twenty-two trials with the same yell calls that I broadcast in areas selected "at random" in the forest (i.e., away from baits). Therefore vocal recruitment cannot by itself explain the gatherings of ravens at baits. It follows that other mechanisms are also involved. The following points are relevant:

18. Except for one special circumstance (near dusk, near a deer kill, shortly after the discovery of the bait by one raven), the main crowds came in groups in one or more straight lines of flight at or before dawn. Since the birds sleep in crowds, they probably came directly from where they spent the night.

19. Birds must have been following each other, because many more came at dawn than had ever been at the bait before.

20. The dawn arrivals often came as a group (up to fifty-two individuals) on consecutive dawns from the same direction where there was a known roost of a similar number of birds, some of which had been captured and marked at the bait site. Therefore, the main recruitment is from the nocturnal communal roost even though these roosts are often very ephemeral in any one location.

21. If the bait was fresh and ample the day before, the dawn arrivals came early in a tight, large, very noisy group, which could be heard giving recruitment yells and other vocalizations while they were well over a mile away, whereas if they were coming to already depleted bait, they came later, in smaller groups, and they were almost silent. Perhaps the recruits simply follow early-leaving noisy birds. If so, leaving early and being noisy would be an effective mechanism of "active" recruitment (because the alternative is to discourage following by concealing knowledge of one's find by delaying departure and remaining silent). However, the connection between food and roosts was even more intimate; communal roosts are often formed very close to the food bonanza itself.

Observations 18–21 show that the *long*-range recruitment is from or through communal roosts. Some preliminary observations are suggestive, but the mechanism for recruitment from the roost remains so far uninvestigated, and I next sought to answer the, to me, much more interesting question of why the ravens recruited others to share their food bonanzas.

The following hypotheses were considered to try to explain why ravens actively recruit:

22. *Improved vigilance for predators.* Perhaps ravens sought a crowd to have "more eyes" to be alert to danger. Argument for: The juveniles are fearful at baits. Arguments against: (1) From my own observations and from the ample literature, it is clear that ravens have little to fear from even the largest hawks. Therefore, according to this hypothesis, it is the vulnerable jays and crows who should recruit instead. (2) If ravens fear the mammalian ground predators instead of hawks, they should prefer to feed at baits high up in trees, rather than those on the ground. My experiment refuted this.

23. *Calling in carcass openers.* My ravens were unable to feed at unopened carcasses, and the recruitment may be meant for carnivores to open up carcasses so that the birds can feed, with other ravens eavesdropping. This hypothesis is totally unsatisfactory for a variety of reasons. The most obvious is that no recruitment occurred at unopened carcasses, but recruitment was often almost immediate if, after weeks or months, the carcass was finally opened, provided ravens already knew of the site.

24. *Sharing the danger of the bait.* To a carcass specialist the bait could be either a sleeping animal, which could lash out in self-defense, or it could be something safe to feed from. Given the perceived danger, a bird could recruit others, or at least another, to test the bait, and later join it or chase it away. The main argument for this hypothesis is that naive birds seem to have an exaggerated fear of anything new. But arguing against it is the fact that I witnessed only one instance where a crowd was assembled before feeding began. In all other cases that I am aware of, the major recruitment occurred *after* feeding began. Therefore, this hypothesis is not the prime mover for recruitment behavior, although it could be a contributing factor.

25. *Food preparation and maintenance.* Snowstorms are frequent, and many birds working at a bait together may

help to keep it shoveled out. A number of birds could also aid one another in tearing the hide. This hypothesis is not tenable. There is *less* recruitment rather than more during heavy snowfalls, and birds were never observed to cooperate in tearing or other food maintenance.

26. *Keeping competing carnivores away.* One coyote alone has easy access to a bait with fifty ravens; ravens always give way. Therefore, it makes no sense to recruit for this reason.

27. *Reciprocal altruism.* According to this hypothesis, a raven does a favor to another one it knows, because it expects that one to repay the kindness sometime in the future. Accordingly, it would not show a food bonanza to another unless there was a reasonable chance of its being reciprocated. But within six days of marking, only thirty-six of sixty-one identified birds remained at a bait. After four weeks, only seven remained, and after six, only one was left, while new ones had taken their place. For over three years now, most of the birds marked in any one area one year are gone the next. Composition of feeding crowds changes from hour to hour. These and many more data showed that there is free movement between crowds. Therefore, the "groups" are not stable, making this hypothesis unlikely.

28. *"Hopeful" reciprocity.* If the cost of sharing is extremely low (such as if the recruited birds eat very little of the pile and/or the attending carnivores eat most of the meat anyway), and the benefits of being reciprocated are very high (saving the bird's life when it needed food), then perhaps sharing with an individual it does not know it will ever meet again is potentially advantageous. Although I at first advanced this hypothesis as a strong contender, I am doubtful now because, as it turned out, the cost of sharing is probably high: Those ravens who shared lost even the largest food bonanzas (a deer or sheep was eaten in two days, a moose in a week), whereas those who did *not* share had up to months of ample feeding at a time when little other food was available. Nevertheless, under more natural conditions carnivores are likely to reduce greatly the time that guarded meat remains available.

29. *Kin selection.* According to this hypothesis, ravens would share with their relatives and discriminate against

nonrelatives. This hypothesis could potentially apply, like reciprocal altruism, where animals live together in extended family groups, or where they breed together in the same area. Ravens do neither. According to detailed studies made in Great Britain, northern Germany, and the northwestern United States, young ravens disperse from their natal area by the first winter, often moving hundreds of miles away. Numerous other young from unknown destinations take their places. My own observations of thirteen locally raised young and ninety-two other nonbreeders entirely confirm this movement. Furthermore, although dominant birds seem to stay longer, turnover of juveniles at any one bait is very rapid. The dozens of studies of the common raven throughout the world all show that the breeding birds are territorial, nest well apart from others, and do not normally associate with neighbors. It is possible that kin selection may have at one time allowed sharing behavior to start evolving, but it is almost certainly not of major significance now.

30. *Status enhancement.* Perhaps a raven that shares is recognized by its peers and thereby gains "friends" or a friend from whom it is able to choose a mate later. This and the next hypothesis, with which it is interrelated, will be discussed after I present additional evidence.

31. *Juvenile gangs.* Although mated and dominant adult ravens live in permanent domains, the low-status juveniles leave the natal area in the fall and are forced to wander. Do they join forces in ephemeral bands to overpower other juveniles or adults, to take over baits?

Given the evidence of my experiments and observations as well as the solid results of other researchers on ravens all over the northern hemisphere, I conclude that the first eight hypotheses (22–29) do not make much sense as the prime movers for explaining the ravens' sharing behavior. One or another, past or present, may have or have had some effect in *facilitating* the evolution of the behavior by some other mechanism. (Selection depends on the total of the selective pressures.) Various ones are likely to apply simultaneously, but to varying degrees. But by themselves they do not explain it now.

The thrust of my efforts should be in providing evidence that shows something positive, rather than in merely narrowing the possibilities. The following information is therefore meant to illuminate the last two hypotheses.

32. Eighty-two out of ninety-one crowd-feeders (from four separate crowds) were nonbreeding juveniles or subadults. (The literature had indicated that the crowd-feeders are juveniles, but these data are the first real proof.)

33. The juveniles and subadults were highly vagrant, while some of the adults were resident year-round. (Both these results are a confirmation of well-known phenomena about ravens already documented by many others, and therefore I won't detail my own evidence here.)

34. The yelling (recruitment calls) were never given by pairs at baits in hundreds of hours of watching pairs only.

35. None of twenty-five discoveries of baits (by individuals or by pairs) was accompanied by yells. Recruitment yells came only later after a number of birds had gathered. They were then conspicuous in all crowds, but given only by very few members of that crowd.

36. Yells were given (probably by the most dominant juveniles) just before feeding began when many birds were already in the vicinity. Since the yell was given only when other birds were near, it acted as a "rallying cry" to get birds to go down to feed together. (I am not referring to motivation here, only to evolutionary mechanisms.)

37. Those ravens that first went down to the food alone were often attacked by the resident adults. As seen from aviary experiments, although the adults were relatively unafraid of the bait (in sharp contrast to the juveniles, which feared almost all baits), they often guarded it for days or weeks, without feeding, thus *seeming* to be afraid of it because they did not go down to feed to "lead" the juveniles in.

38. Mated resident adults vigorously chased other ravens near the bait, but when there were large crowds of juveniles and such efforts became futile, then they stopped chasing and fed.

39. The resident adults in the presence of the vagrants showed feather posture signifying dominant status, while the juveniles near them showed submissive postures.

40. By themselves, both resident pairs and juveniles showed neutral feather postures.

41. No resident adults were ever excluded by juveniles, but residents sometimes guarded baits daily for several weeks, successfully excluding all others.

42. If recruitment is for sharing the wealth, then the larger the food bonanza, the more recruitment there should be. But crowd size was independent of bait size, which ranged from 20 to nearly 1,000 pounds. Therefore, recruitment is more probably related to gaining or maintaining *access* to the food than to sharing the wealth. This would explain the fighting commonly observed after feeding began.

43. In five different experiments over two years, a total of about ten baits were (each time) simultaneously provided in areas where no recruitment had been observed sometimes for months, even though raven pairs were seen at the baits. Presumably no pair could defend ten baits simultaneously, and in all five experiments at least one bait pile was recruited to *within one day* and almost all were then eaten by crowds, one after another. Therefore, neither the site nor the kind of bait, as such, is what limited recruitment. It seems that normally with single baits it is the presence of resident adult pairs that delays access by roving juveniles.

44. Two patterns of bait use were seen. Many baits were visited *only* by a pair (up to two winters in one case), or they were visited by a crowd.

Conclusion: The recruitment is by vagrant juveniles (27, 32–34) which then gain or maintain access to food otherwise defended by resident pairs of adults (34–44).

The adults never yell at baits because that would attract juvenile scouts. In turn, the juveniles don't yell when they *discover* baits because that would attract the adult pairs. Recruitment occurs whether or not the residents are there initially, because they will probably be alerted in any case. The yells function as a rallying cry when the juveniles can safely give away their presence because they have assembled a critical number from some distant source. The resident adults are dominant and chase the juveniles that are not in crowds, but they quit defending the bait after they are overwhelmed, because chasing would then be a waste of their time and efforts.

There is one other tantalizing set of observations that suggests a second reason for recruitment. If recruitment is *only* to gain or maintain access to the food, then it should soon stop after the juveniles can "safely" feed at the bait. Instead, numbers often continued to increase. Undoubtedly one reason for this is that the increased commotion attracts even more birds. But there may be still more to it.

The following set of ideas and observations suggests (but does not prove) that recruitment could be related to sexual selection, where mate choice is related to ability to find or gain access to food.

45. Female ravens depend utterly on their mates to feed them and their small young for over a month every year.

46. In many birds, part of the mate-choice ceremonies during courting involves the male providing food for the female.

47. Excellent evidence from many researchers shows that whether or not a raven can exist and breed in a domain depends on competition with other ravens. There are dominance contests at baits. It would therefore be advantageous for a raven to have a dominant individual as a mate.

48. Since ravens choose one mate for a lifetime, and may live over forty years, they are predictably very choosy about potential mates.

49. Detailed previous research has clearly shown both that juvenile males fight for dominance among themselves and that the more dominant ones are preferred by the females.

50. The juveniles that yell (recruit) are probably the most dominant birds, and the recruiters win contests (and status?).

51. Ravens can begin to court when only about six months old (in the first fall) but mating does not take place until they are at least three years old. Courting is year-round in unmated dominant birds, but low-status birds are reproductively suppressed by dominants.

52. Young ravens fear all strange food objects and learn *not* to fear only by experience. My work with captive birds confirmed previous studies showing that "leaders" emerged from the crowd and led others to feared food. The leaders were in turn courted, while those who did not lead (who

were not brave) were not courted. Thus, those who directly or indirectly provide food or have the most access to it are the most "popular" or powerful.

53. Almost all raven researchers confirm that crowd birds, which are primarily juveniles, often pair up.

How does this information add up? So far, I see only one major picture or pattern when I try to fit these more than fifty widely disparate sets of data together (the number of alternative patterns that I see is in inverse proportion to the number of facts I simultaneously keep in mind). According to this picture, the young ravens leave home to wander. They are gregarious, joining other juveniles to roost and feast with, and to find an attractive mate. An unmated raven finding food invites eligible singles to join him (or her?) at the feast, thereby not only gaining or maintaining access to the food, but possibly also increasing its status and demonstrating fitness as a future provider for rearing offspring. It is an elegant, simple, and beautiful system. But it is clothed by intricate detail and subtlety. As far as I know, no other animal shows a similar system. However, sometimes when I am fanciful and envision ravens studying humans, I can't help but wonder what they would make of some of our customs, and how *they* might arrive at scientific conclusions about them.

APPENDIX

The following data are compiled from the whole study to show overviews of the information collected. Some of these data sets give key evidence. Others illustrate a problem confronted in doing the research; they illuminate a small portion of the overall problem I was trying to solve, but at the same time they give a false picture of the whole if the small part is extrapolated to the whole. For example, Figure 19 shows that the amount of fighting among ravens at a sheep carcass was a direct function of crowd density. That was true, at that time, in that place, and I expect it will be as true when one observes it at other baits as well. But it hides the very important fact that one or two ravens will fight a third that may not be crowding at all (such as territorial birds against an intruder in the carcass vicinity but not directly on it).

Near the beginning of the study I was often lured astray by selective data. Information can also be disinformation, depending on the context. Therefore, the evidence has to be evaluated in total, to see how it fits together, because each has a place in the picture but is valueless by itself.

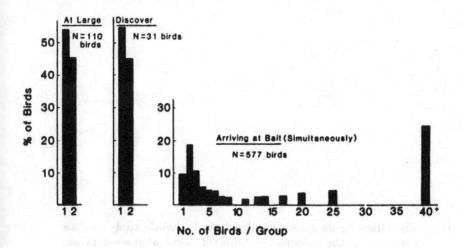

Fig. 1. These data lump observations collected over 3 autumns and 3 winters. They show that the 110 ravens I saw in Vermont, New Hampshire, and Maine (at the study site when no bait was present, and while driving to and from it) were exclusively flying as single birds or as 2 birds flying together. The discoverers at baits were in nearly identical proportion singles or doubles, but after bait discovery the percentage of birds arriving as singles or as doubles was only 30%; 70% of the arrivals were now in groups. These data indicate that the ravens forage in the north woods of New England by flying singly or in pairs, and that either others follow or they lead in others after a bait discovery has been made.

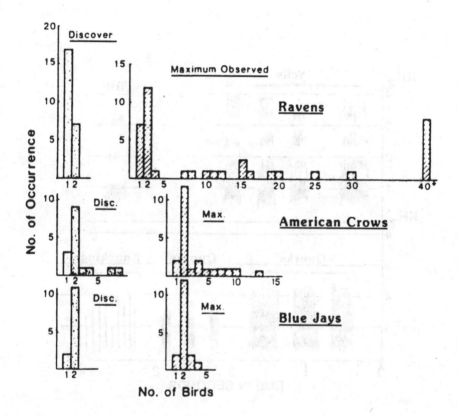

Fig. 2. The data here show several things. First, as in the previous figure, we see that most baits are discovered by ravens arriving as single birds. Only subsequently do large numbers utilize the bait. The pattern with crows and blue jays show that most discoveries are made by *pairs*, and that *pairs* are then the most common utilizers of the baits. Thirdly, the most obvious difference between the ravens and the other 2 corvids is that only with the ravens do we see accumulations of birds above and beyond the numbers of birds that make the bait discovery. This comparative data suggests that the raven accumulations at the bait are not solely a result of a passive target provided by the birds already there, or the patterns should be similar in all 3 species.

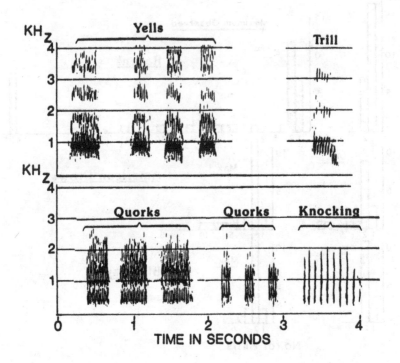

FIG. 3. Sonograms of common vocalizations near the bait. The juvenile yells are given prior to and during feeding when large numbers of birds are present, but not when only pairs are present. The trills are also associated with food, but also with other excitement. The long quorks are given by the resident adult birds when challengers are near. Presumably they function as territorial advertisement. There is a great variety of quorks, however, and they undoubtedly convey various meanings. The knocking is given by females.

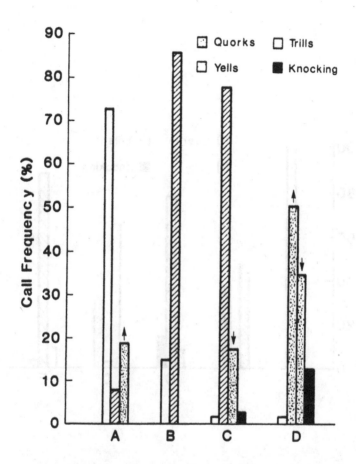

FIG. 4. Vocalizations at an opened sheep. $A =$ Vocalizations ($N = 63$) 10 minutes before feeding began. $B =$ Vocalizations ($N = 20$) during the 10 minutes after feeding began. $C =$ Vocalizations ($N = 110$) during the next 6 hours. $D =$ Vocalizations ($N = 203$) during 3.5 hours after one bird had seen me, and when the birds (at least 4 present) no longer came down to the carcass. The different bars designate different types of calls. Arrow pointing up = relatively high-pitched quorks. Arrow pointing down = lower-pitched quorks. (A similar pattern of predominance of trills prior to feeding, yells during feeding, and quorks and knocking shortly after the birds were scared off the bait was seen also at a landfill.)

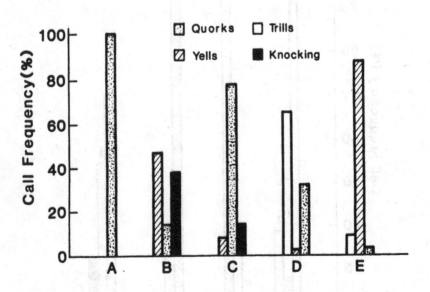

Fig. 5. Vocalizations as a function of feeding sequence at a large bait (approximately 90 lbs. of beef suet). A = Discovery by a single bird (all vocalizations for 10 minutes after discovery, $N = 55$). $B = 81$ minutes after the bird left, and the simultaneous arrival of four (juveniles?) birds (all vocalizations for the first 10 minutes, $N = 49$). C = all vocalizations for the subsequent 3 hours ($N = 76$), (possibly both the territorial defender and the juveniles?) D = Vocalizations ($N = 46$) by 1 of the 4 birds during approximately 12 minutes just before 5 birds began feeding. E = Vocalizations ($N = 88$) during 10 minutes, 6 hours after the juveniles swamped the bait and began feeding with 15 birds present at the same time.

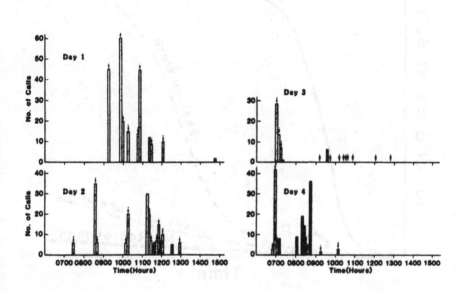

FIG. 6. Vocalization of a presumed pair after discovery of a cow carcass (Day 1) and over the next 3 days. Feeding occurred only on the third day at arrows. Only quorks (stippled) and knocking (filled) was heard. Period over bar = short quorks. Short horizontal line over bar = long quorks. Arrow pointing down = deep quorks. Arrow pointing up = high quorks.

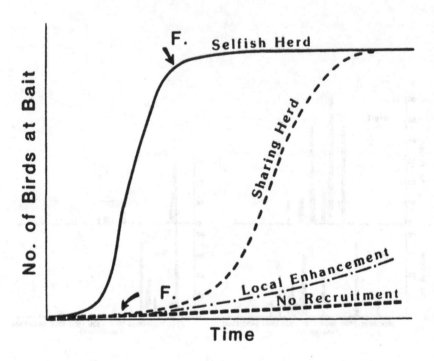

FIG. 7. Expected *patterns* of buildup of birds at a food bonanza depending on different contingencies. If there is no recruitment, then numbers should increase gradually, as new birds find the bait. If, after finding the bait, the birds already there provide a larger target to see or follow, one might expect a gradually increased rate of buildup. The local enhancement would be due to a "parasitic" effect where successful birds unwillingly or passively give away the food location. The last two curves show expected increases in bird numbers as related to the start of feeding (*F*) during *active* recruitment. If the birds are recruiting to share the risk associated with their fear of the bait, they should recruit *first* and *then* feed, as well as stop recruiting after they no longer fear the bait. On the other hand, if they share the food and not the risk associated with it, then they should test it first, and then after finding that it is suitable and no or little risk is associated with feeding, they should bring in others that might share the benefits.

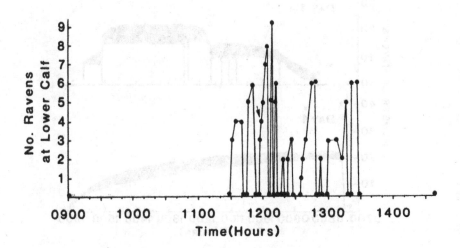

FIG. 8. This shows the number of ravens feeding at a calf carcass, with simultaneous flight from the carcass and simultaneous return. Birds did not begin to feed until the third descent to the carcass (at arrow). Such data are consistent with the "selfish herd" hypothesis, where the animals try to share the apparent risk they perceive at the food.

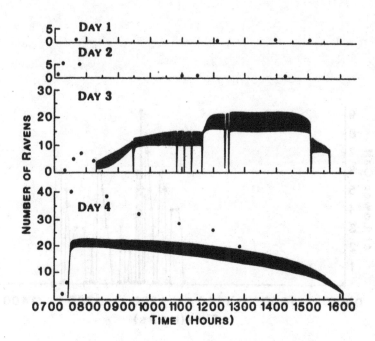

Fig. 9. The number of ravens seen near an opened sheep carcass the day when it was discovered (Day 1), when feeding began (Day 3), and the following day until it was consumed. Filled areas indicate the range in raven numbers feeding at the carcass at any one time (exact numbers varied on a minute-by-minute basis). Vertical gaps show when all birds left, and again quickly returned. Note 5 or more birds coming at dawn the second and third day after discovery, and 40 the fourth day. This sequence shows the rapid buildup of birds often seen in a short time, often after no birds had been seen for several days. Since most of the increase in bird numbers here occurred only *after* feeding began, these data are more consistent with sharing behavior, rather than with the selfish herd hypothesis.

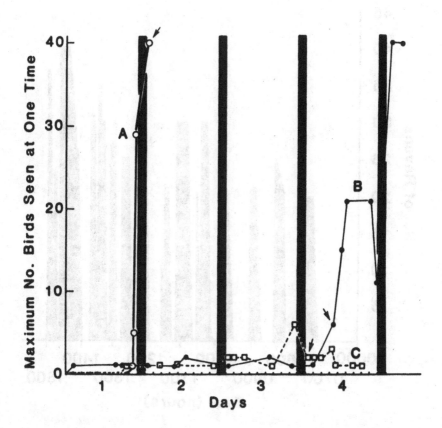

FIG. 10. The maximum numbers of ravens seen at any one time after bait discovery (Day 1) at three different baits. (Nights are shown as thick vertical bars.) This figure shows three different responses. *A* shows fast recruitment on one evening, and feeding in the next morning. *B* shows no feeding and recruitment until three days after bait discovery. *C* also shows feeding only after the third day of bait discovery, but in this case there was no recruitment.

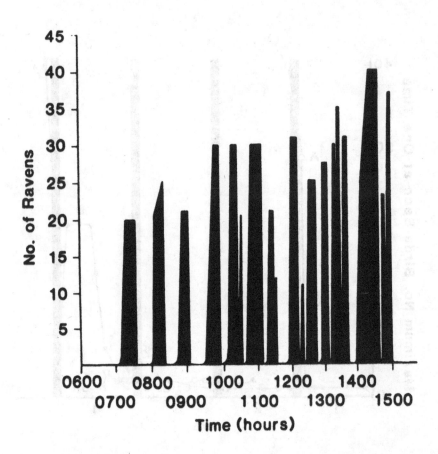

FIG. 11. Number of ravens feeding (Dec. 9, 1986) at a bait, on the second day after feeding (by about 10 birds) started. This graph shows the birds descending to and leaving the bait more or less synchronously, with up to half-hour periods of no ravens alternating with times when 20–40 were feeding simultaneously. When the ravens were off the bait, it was fed from either by 1–4 blue jays or 2 crows. These data suggest that the bait is not the only thing they fear.

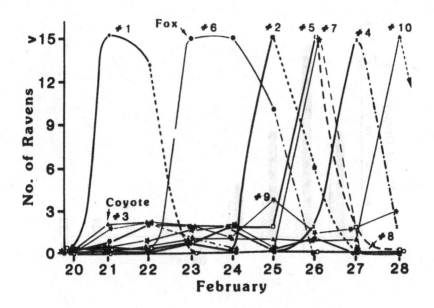

FIG. 12. Consecutive rapid buildup of numbers of ravens at 7 of
the 10 baits (< 35 lbs. of meat each) all put out on the same
day, Feb. 20, 1987. One bait (#3) was consumed by a coyote.
One (#8) was undiscovered. One (#6) was visited by a fox,
which delayed depletion of the bait by the ravens. Most baits
were consumed almost entirely by the ravens over 1 day. Dashed
lines indicate that the meat was gone before the time of the next
visit. These data are inconsistent with birds finding baits independently of one another.

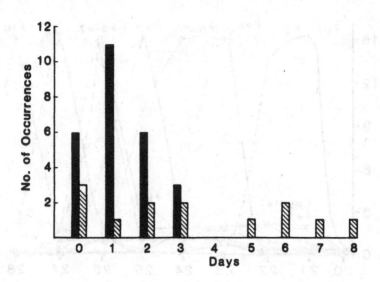

FIG. 13. Number of days after 27 baits were put out to when they were discovered (solid bars) and to when they were recruited to (slashed bars) and then consumed by crowds.

We here see that ravens discovered all 27 baits of this specific experiment in 3 days, but they did not recruit to some of them until after a week. More significantly, recruitment could occur the *same* day of bait discovery. If lack of recruitment (commonly observed for months) were due to fear of baits alone, then there should have been no recruitment following discovery, and then gradual *increase* in recruitment over subsequent days. It is important to note that this experiment (a composite of three) involved saturating an area with *many* baits, so that presumed territorial defense by pairs would not have been possible because 2 birds could not patrol a dozen or more baits spread out over several miles simultaneously.

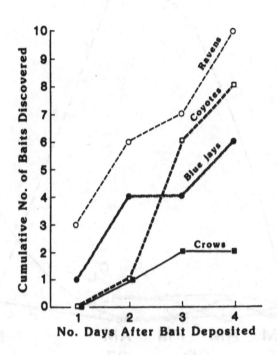

FIG. 14. Cumulative number of baits (20–25 lbs. meat) out of 12 available (in an experiment conducted Nov. 4–8) that were discovered by ravens, coyotes, blue jays, and crows, as determined by tracks in the snow and visual sightings during the first 4 days after the bait was deposited. All baits were within 4 miles of one another.

This shows that the lack of crow or blue jay crowds cannot be accounted for by these animals' not discovering the baits. Therefore, what distinguishes the use of baits by raven crowds must be related to a factor occurring after discovery.

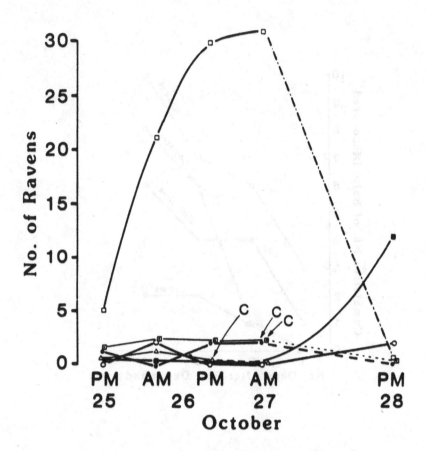

Fig. 15. Recruitment can be almost "instant" when many baits are simultaneously put out in one area of a few square miles. This figure shows maximum number of ravens observed at 6 different baits put out in the afternoon of Oct. 25, 1986, that were discovered by ravens. C = visited by coyote (determined by tracks on snow). Dashed lines indicate depletion of the bait. All baits were within 8 air miles of one another. Note that at some baits there was no recruitment, even though one or two ravens visited the bait.

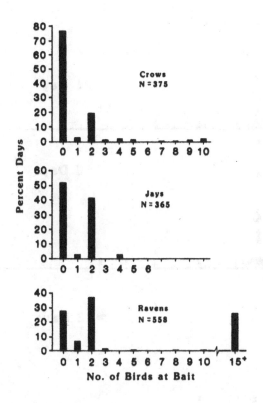

Fig. 16. This figure emphasizes that the percent of days that a bait does not have one of the corvids is lowest for ravens (about 30%); that is, they monopolize the baits. Secondly, we see that the pattern of bait use by *pairs*, as opposed to other combination of numbers, is nearly as strong with ravens as it is with jays when the total *time* that bait use is considered. However, about 30% of the days that a bait is available it is being visited by large numbers of ravens, unlike the other 2 corvids. This figure, showing the amount of time that various groupings of birds utilize a bait greatly underestimates the bait *use* as a function of bird groupings; a large grouping of ravens commonly consumed a 30-lb. pile of meat in 1 day, whereas a pair might be the exclusive users for more than a month. (The N refers to days, and since the monitoring refers to all 3 corvid species at the *same* baits, the N should be equal, except that only ravens were counted for the first part of the study.)

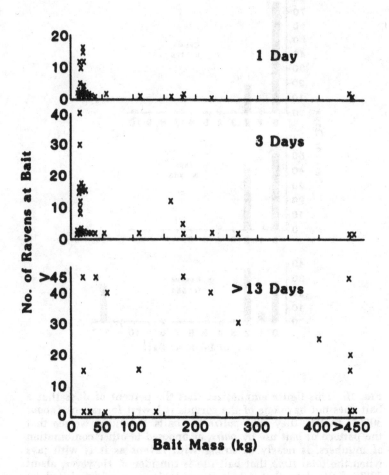

Fig. 17. Number of ravens at differently sized baits after 1, 3, or more than 13 days. These data show that large numbers of ravens can gather at a bait even in 1 day, but that the total numbers of birds at baits ranging from 30 to over 1,000 lbs. is unrelated to bait size, at least in the first 2 weeks that the bait is available. (The apparently greater initial accumulation of numbers at the smaller baits is due to greater overall numbers of small-sized baits.)

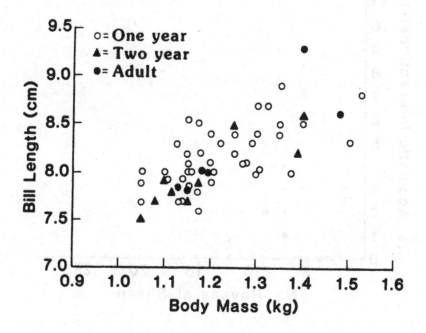

FIG. 18. Adults cannot be distinguished by larger size. We see here bill lengths and body mass of ravens captured at 2 different occasions in Jan.–Feb. 1987, in Weld, Me.

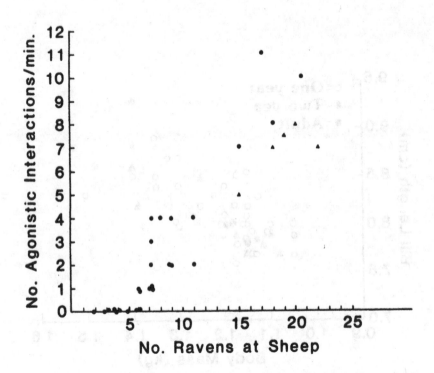

Fig. 19. One of the costs of feeding in crowds by ravens is increased squabbling, especially as the amount of bait declines. Shown here is the number of agonistic interactions as a function of number of ravens feeding at or surrounding a frozen sheep carcass, on Jan. 4 and 5, 1986. Triangles: First day of feeding. Solid circles: Second day of feeding when most of the meat was consumed.

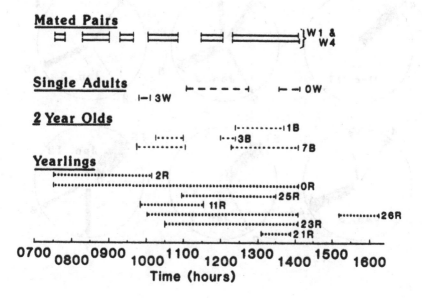

FIG. 20. Arrival and departures of 1 mated pair, 2 single adults, 3 2-year-olds, and 7 yearlings individually marked at a bait on Jan. 25, 1987. Only the mated pair (who showed conspicuous territorial behavior) came and went together.

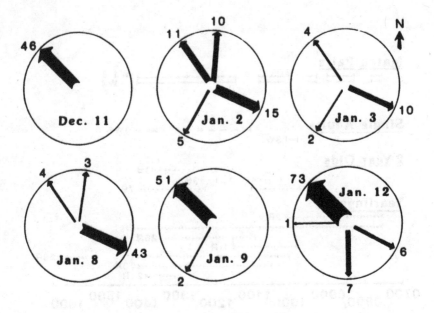

FIG. 21. Numbers of ravens and the directions of their arrival as I saw them from the top of a tall spruce tree from 0620 to 0700 h (sunrise was at 0720 h) at a large bait on 6 different days. Note that up to 73 birds came from the NW, while 6–43 came on 4 days from the SE.

These large numbers of birds (coming before sunrise) from the same direction on different days suggest arrival from a communal nocturnal roost. But birds come from more than one roost. A known roost was located 15 miles to the NW.

NOTES

AN INTRODUCTION

18 *and other names:*
R. Ayre, 1963. *Sketco, The Raven.* Toronto, Macmillan.

18 *Corvidae should include birds of paradise:*
D. Amadon, 1944. The genera of Corvidae and their relationships. *Am. Mus. No.* 1251:1–21.

19 *proteins more similar to those of shrikes:*
C. G. Sibley, 1970. A comparative study of the eggwhite proteins of passerine birds. *Peabody Mus. Nat. Hist. Bull.* 32:131, p. Yale Univ.

19 *Sibley and his colleagues:*
C. G. Sibley, J. E. Ahlquist, and B. L. Monroe, Jr., 1988. A classification of the living birds of the world based on DNA-DNA hybridization studies. *Auk* 105:409–23.

20 *Check-list of Birds of the World:*
E. Mayr and J. C. Greenway, Jr., 1962. *Checklist of Birds of the World,* Vol. 15. Cambridge, Mass., *Mus. of Compar. Zool.*

20 *Malcolm Jollie at Northern Illinois University:*
M. Jollie, 1978. Phylogeny of the species of *Corvus. The Biologist* 60:73–108.

20 *measurements accumulated by George Willett:*
G. Willett, 1941. Variation in North American ravens. *Auk* 58: 246–49.

20 *males averaged 1.38 kilograms:*
C. B. White and T. J. Cade, 1971. Cliff-nesting raptors and ravens along the Colville River in arctic Alaska. *Living Bird* 10:107–150.

24 *in his 1983 book on the Koyukon natives:*
R. K. Nelson, 1983. *Make Prayers to the Raven: A Koyukon View of the Northern Forest.* Chicago, Univ. of Chicago Press.

26 *during his archeological excavations:*
H. B. Collins, 1966, in A. Stefferud, ed., *Birds in Our Lives.*
Washington, D.C., U.S. Dept. Interior, U.S. Govt. Printing Office,
pp. 279–83.

26 *a pioneer raven researcher in Germany:*
J. Gothe, 1961. Zur Ausbreitung und zum Fortpflanzungsverhalten
des Kolkraben (*Corvus corax* L.) unter besonderer Berücksichtigung der Verhältnisse in Mecklenburg, in H. Schildmacher,
Beiträge zur Kenntnis deutscher Vögel. Jena, Fischer, pp. 63–129.

27 *raven persecution in the West:*
J. R. Mead, 1986. *Hunting and Trading on the Great Plains:
1859–1875.* Norman, University of Oklahoma Press.

29 *the Raven has almost forsaken:*
T. Nuttall, 1903. *Birds of the United States and Canada.* Boston,
Little, Brown, and Company.

29 *concluded that the raven was "uncommon or rare":*
A. C. Bent, 1964. *Life Histories of North American Jays, Crows,
and Titmice.* Part I. New York, Dover.

29 *Barrows wrote that in Michigan:*
W. B. Barrows, 1912. *Michigan Bird Life.*

32 *crows and corvids are found:*
H. Bell, 1984. The house crow is coming. *R.A.U.U. Newsletter*
60:5.

32 *in the Saskatchewan prairies:*
C. S. Houston, 1977. Changing patterns of Corvidae on the
prairies. *Blue jay* 35:149–56.

32 *interstate highway overpasses and billboards:*
C. M. White and M. Tanner-White, 1988. Use of interstate high-
way overpasses and billboards for nesting by the common raven
(*Corvus corax*). *Great Basin Naturalist* 48:64–67.

33 *Vermont, Maine, and New Hampshire:*
S. B. Laughlin and D. P. Kibbe, eds., 1985. *The Atlas of Breed-
ing Birds of Vermont.* Hanover, NH, University Press of New
England.

37 *as studied by the zoologists:*
G. E. Woolfenden, 1975. Florida scrub jay helpers at the nest.
Auk 92:1–15; idem. and J. W. Fitzpatrick, 1984. *The Florida
Scrub Jay: Demography of a Cooperative-breeding Bird.* Prince-
ton, N.J., Princeton Univ. Press.

37 *as demonstrated:*
J. L. Brown, 1970. Cooperative breeding and altruistic behaviour
in the Mexican jay, *Aphelocoma ultramarina. Anim. Behav.* 18:
366–78; idem, 1972. Communal feeding of nestlings in the Mexican jay (*Aphelocoma ultramarina*): Inter flock comparisons.

Anim. Behav. 20:395–403; idem, 1987. *Helping and Communal Breeding in Birds: Ecology and Evolution.* Princeton, N.J., Princeton Univ. Press.

37 *found in the pinyon jay:*
R. P. Balda and G. C. Bateman, 1971. Flocking and annual cycle of the pinon jay, *Gymnorhinus cyanocephalus. Condor* 73:287–302.
J. M. Marzluff and R. P. Balda, 1987. Resource and climatic variability: influence on sociality of two southwestern corvids, in C. N. Slobodchikoff, ed., *The Ecology of Social Behavior,* New York, Academic Press, pp. 255–83.

38 *come upon a bonanza:*
B. Heinrich, 1987. *One Man's Owl.* Princeton, N.J., Princeton Univ. Press.

38 *may be modified by evolution:*
S. B. Vander Wall and R. P. Balda, 1981. Ecology and evolution of food storage behavior in conifer-seed-caching corvids. *Z. Tierpsychol.* 56:217–42.

39 *produce a sticky saliva:*
W. J. Bock, 1961. Salivary glands in the gray jay (*Perisoreus*). *Auk* 78:355–65.

39 *deep snows cannot obliterate them:*
D. D. Dow, 1965. The role of saliva in food storage by the gray jay. *Auk:* 82:139–54.

39 *dervied from cached food:*
S. B. Vander Wall and R. P. Balda, 1977. Coadaptation of the Clarke's nutcracker and pinyon pine for efficient seed harvest and dispersal. *Ecol. Monogr.* 47:89–111.
D. F. Tomback, 1977. Foraging strategies of Clarke's Nutcracker. *The Living Bird* 16:123–58.

RAVENS AS HUNTERS AND SCAVENGERS

49 *some of the available sources:*
A. L. Nelson, 1934. Some early summer food preferences of the American raven in southeastern Oregon. *Condor* 36:10–15.
S. B. Tyrell, 1945. A study of the northern raven. *Auk* 62:1–7.
A. C. Bent, 1964. *Life Histories of North American Jays, Crows, and Titmice.* Part I. New York, Dover.
S. A. Temple, 1974. Winter food habits of ravens on the arctic slope of Alaska. *Arctic* 27:41–46.
J. L. Dorn, 1972. "The common raven in Jackson Hole, Wyoming," M.S. Thesis. Univ. of Wyoming, Laramie, Wyo.
K. A. Engel, L. S. Young, W. G. Kell, and A. J. Brody, 1987. Implications of communal roosting by common ravens to opera-

tion and maintenance of the Malin to Midpoint 500 KV transmission line, in K. Steenhof, ed., Snake River Birds of Prey Report, Boise, Idaho, BLM, U.S. Dept. of Interior, pp. 34–55.

49 *eggs in its stomach:*
G. F. Knowlton, 1943. Raven eats Mormon cricket eggs. *Auk* 60:273.

50 *pellet analysis shows:*
R. Hoffman, 1920. A raven pellet. *Auk* 37:453–54.
S. B. Tyrell, 1945. A study of the northern raven. *Auk* 62:1–7.
S. A. Temple, 1974. Winter food habits of ravens on the arctic slope of Alaska. *Arctic* 27:41–46.
R. F. Harlow, R. G. Hooper, R. D. Chamberlain, and G. H. S. Crawford, 1975. Some winter and nesting season foods of the common raven in Virginia. *Auk* 92:298–306.
I. Newton, P. E. Davis, and J. E. Davis, 1982. Ravens and buzzards in relation to sheep-farming and forestry in Wales. *J. Applied Ecol.* 19:681–706.
M. Marquiss and C. J. Booth, 1986. The diet of ravens (*Corvus corax*) in Orkney. *Bird Study* 33:190–95.
P. J. Ewins, J. N. Dymond, and M. Marquiss, 1986. The distribution, breeding and diet of ravens (*Corvus corax*) in Shetland. *Bird Study* 33:110–16.
L. S. Young, K. A. Engel, and A. Brody, 1986. Implications of communal roosting by common ravens to operation and maintenance of the Malin to Midpoint 500 KV transmission line. *Snake River Birds of Prey Research Project*, Annual Report 1986, pp. 36–59.

50 *Temple's analysis of 684 pellets:*
S. A. Temple, 1974. (*Arctic* 27:41–46)

50 *reported seeing a raven:*
F. F. Mallory, 1977. An ingenious hunting behavior in the Common Raven (*Corvus corax*). *Ont. Field Biol.* 31:77.

50 *lived "almost exclusively" on moles:*
R. Hoffman, 1920. (*Auk* 37:453–54)

51 *of the eighty-four stomachs:*
A. L. Nelson, 1934. (*Condor* 36:10–15)

51 *killed a "sickly hen":*
R. McManus, 1935. Feeding habits of the raven in winter. *Auk* 52:89.

51 *from its nest on a cliff:*
V. Marr and R. L. Knight, 1982. Raven predation of feral Rock Dove eggs. *Murrulet* 63:25.

51 *disposed of . . . by the following day:*
S. B. Tyrell, 1945. (*Auk* 62:1–7)

51 *near Chagnon Bay, Alaska:*
D. F. Parmelee and J. M. Parmelee, 1988. Ravens observed killing roosting kittiwakes. *Condor* 90:952.

51 *the eyes of newborn lambs:*
K. H. Larsen and J. H. Dietrich, 1970. Reduction of raven population on lambing grounds with DRC-1339. *J. Wildl. Mgt.* 34: 200–204.

51 *reported in 1925:*
R. C. Ross, 1925. Field notes on the raven. *Condor* 27:172.

52 *ravens attacking live sheep:*
K. Igalffy, 1971. A raven, *Corvus corax*, attacks a sheep. *Larus* 2:11.

52 *and reindeer:*
E. Ostbye, 1969. Raven attacking reindeer. *Fauna* 22:265–66.

52 *describes a pair of ravens hunting:*
C. Maser, 1975. Predation by Common Ravens on feral Rock Doves. *Wilson Bull.* 87:552–53.

52 *"illustrate the ravens' resourcefulness":*
A. C. Bent, 1964. *Life Histories of North American Jays, Crows and Titmice.* Part I. New York, Dover.

53 *watching an otter catch fish:*
R. K. Nelson, 1983. *Make Prayers to the Raven: A Koyukon View of the Northern Forest.* Chicago, Univ. of Chicago Press.

53 *pirating kills from gyrfalcons:*
C. B. White and T. J. Cade, 1971. Cliff-nesting raptors and ravens along the Colville River in arctic Alaska. *Living Bird* 10:107–50.

53 *. . . and owls:*
M. Marquiss, I. Newton, and D. A. Ratcliffe, 1978. The decline of the raven (*Corvus corax*) in relation to afforestation in southern Scotland and northern England. *J. Appl. Ecol.* 15:125–44.

53 *an apparent hermit:*
F. Zirrer, 1945. The Raven. *Passenger Pigeon* 7:61–67.

53 *through age and experience:*
E. K. Dunn, 1972. Effect of age on the fishing ability of Sandwich Terns *Sterna sandvicencis*. *Ibis* 114:360–66.
F. G. Buckley and P. A. Buckley, 1974. Comparative feeding ecology of wintering adult and juvenile Royal Terns (Aves: Laridae, Sterninae). *Ecology* 55:1053–63.
A. Ingolfson and J. T. Estrella, 1978. The development of behavior in Herring Gulls. *Auk* 95:577–79.

54 *Canadian wildlife researcher:*
L. N. Carbyn, 1983. Wolf predation on elk in Riding Mountain National Park, Manitoba. *J. Wildl. Mgt.* 47:963–76.

54 *the snow is deep and killing is easy:*
M. L. Wilton, 1986. Scavenging and its possible effects upon predation. *Alces* 22:155–80.

54 *studying summer scavenging:*
A. J. Magoun, 1976. "Summer scavenging activity in northeastern Alaska." A thesis. Univ. of Alaska, Fairbanks.

55 *eat their droppings:*
D. L. Allen, 1979, p. 288. *Wolves of Minong.* Boston, Houghton Mifflin.

55 *elicited responses from ravens:*
F. H. Harrington, 1978. Ravens attracted to wolf howling. *Condor* 80:236–37.

56 *when Prince Maximilian was near:*
D. L. Allen, 1979. ibid.

56 *where there were reindeer:*
J. Franz, 1943. Über Ernährungs- und Tagesrythmus einiger Vögel im arktischen Winter. *J. Ornith.* 91:156.

56 *speculates that ravens can track wolves:*
L. D. Mech, 1970, pp. 279–287. *The Wolf: The Ecology and Behavior of an Endangered Species.* Garden City, New York, Natural History Press.

56 *the polar night of winter:*
J. Franz, 1943. ibid.

56 *describes being told by Eskimos:*
N. Tinbergen, 1958. *Curious Naturalists.* New York, Doubleday, p. 38.

56 *kills in Alaska:*
J. M. Holzworth, 1930. *The Wild Grizzlies of Alaska.* New York, Putnam, p. 370.

56 *. . . and in Yellowstone Park:*
F. C. Craighead, Jr., 1979. *Track of the Grizzly.* San Francisco, Sierra Club Books.

57 *twenty to twenty-six feet:*
K. E. Campbell, Jr., and E. P. Tonni, 1983. Size and locomotion in teratorns (Aves: Teratornithidae). *Auk* 100:390–403.

57 *asphalt deposits of Rancho La Brea:*
Howard, 1962. A comparison of prehistoric avian assemblages from individual pits at Rancho La Brea, California. Los Angeles County Museum contrib. *Sci* 58:1–24.

CALLING IN CARCASS-OPENERS?

63 *honeyguide:*
H. A. Isack and H. U. Reyer, 1989. Honeyguides and honey gatherers: Interspecific communication in a symbiotic relationship. *Science* 243:1343–46.

WHAT IS ACCEPTABLE EVIDENCE?

91 *article on ravens:*
B. Heinrich, 1986. Ravens on my mind. *Audubon:* 88:74–77.

93 *stressed by Sigmund Freud:*
S. Freud, "Recommendations to physicians practicing psychoanalysis," in J. Strachey, ed., 1963, *The Collected Papers of Sigmund Freud*, vol. 2. New York, Basic Books.

ARE RAVENS HAWKS OR DOVES?

106 *formalized the argument in 1974:*
J. Maynard Smith, 1974. The theory of games and the evolution of animal conflict. *J. Theor. Biol.* 47:209–21.

108 *the tit-for-tat strategy:*
L. A. Dugatkin, 1988. Do guppies play tit-for-tat during predator inspection visits? *Behav. Ecol. & Sociobiol.* 23:395–99.

109 *could have a "hopeful reciprocity":*
B. Heinrich, 1988. Foodsharing reciprocity and communication in the raven, *Corvus corax*, in C. Slabodchikoff, ed., *The Ecology of Social Behavior*. New York, Academic Press, pp. 285–311.

RAVEN INTELLIGENCE

112 *"it could only be concluded":*
F. F. Mallory, 1977. An ingenious hunting behavior in the common raven (*Corvus corax*). *Ont. Field Biol.* 31:77.

113 *without reference to snow or tunnels:*
B. Heinrich, 1988. Why do ravens fear their food? *Condor* 90:950–52.

113 *ravens carry nuts and shellfish:*
T. Nuttall, 1903. *Birds of the United States and Canada*. Boston, Little, Brown, and Company.

113 *well-documented for birds such as crows . . . :*
R. Zach, 1979. Shell-dropping: Decision-making and optimal foraging in northwestern crows. Behaviour 68:106–17.

113 *. . . and gulls . . . :*
C. Oldham, 1930. The shell dropping habit of gulls. *Ibis* 6:
239–43.
D. P. Barash, P. Donovan, and R. Myrick, 1975. Clam dropping
behavior of glaucous-winged gulls. *Wilson Bull.* 87:60–64.

113 *. . . which are of lesser intelligence:*
L. Benjamin, 1983. Studies in the learning abilities of brown-
necked ravens and herring gulls: Oddity learning. *Behavior* 11:
173–194.

113 *ravens playing with fir cones:*
A. C. Bent, 1964. *Life Histories of North American Jays, Crows,
and Titmice.* Part I. New York, Dover.

113 *dropping objects onto human nest intruders:*
S. W. Janes, 1976. The apparent use of rocks by a raven in nest
defense. *Condor* 78:409.

113 *onto incubating birds:*
W. A. Montevecchi, 1978. Corvids using objects to displace gulls
from nests. *Condor* 80:349.

114 *makes a . . . case for awareness being beneficial:*
D. Griffin, 1984. *Animal Thinking.* Cambridge, Mass., Harvard
Univ. Press, pp. 120–21.

114 *detailed observations of behavioral ecologist Eberhard
        Gwinner:*
E. Gwinner, 1965. Beobachtungen über Nestbau und Brutpflege
des Kolkrabens (*Corvus corax*) in Gefangenschaft. *J. Ornithol.*
106:145–78.

114 *pry a peanut out of a crack:*
S. G. Jewett, 1924. An intelligent crow. *Condor* 26:72.

114 *using tools to retrieve food:*
T. B. Jones and A. C. Kamil, 1973. Toolmaking and tool-using
in the northern blue jay. *Science* 180:1076–78.

114 *to probe into a hollow stem:*
R. I. Orenstein, 1972. Tool-use by the New Caledonian crow
(*Corvus monaduloides*). *Auk* 89:674–76.

114 *to retain water to drink:*
J. B. Reid, 1982. Tool-use by a rook and its causation. *Anim.
Behav.* 30:1212–16.

115 *using automobiles as "nutcrackers":*
T. Maple, 1974. Do crows use automobiles as nutcrackers? *Western Birds* 5:97–98.

115 *acorns to hammer, as "anvils":*
J. R. Michener, 1945. Californian jays, their storage and recovery of food and observations at one nest. *Condor* 47:206–10.

115 *piegons, rats, and monkeys:*
R. W. Powell, 1972. Operant conditioning in the common crow (*Corvus brachyrhynchos*). *Auk* 89:738–42.

TAME BIRDS FROM THE NEST

140 *great differences in play behavior . . . :*
E. Gwinner, 1966. Über einige Bewegungsspiele des Kolkrabens (*Corvus corax* L.) *Z. Tierpsychol.* 23:28–36.

141 *. . . and in nest-building:*
E. Gwinner, 1965. Beobachtungen über Nestbau und Brutpflege des Kolkrabens (*Corvus corax*) in Gefangenschaft. *J. Ornithol.* 106:145–78.

TERRITORIAL ADULTS AND WANDERING JUVENILES

153 *holding a permanent territory, repeated mention of "flocks" of juveniles:*
D. Goodwin, 1986. *Crows of the World,* 2nd ed. Seattle, Univ. of Wash. Press.

153 *resides year-round in its territory:*
R. L. Knight and M. C. Call, 1974. The common raven. U.S. Dept. Interior Tech. Note No. 344.
I. Newton, P. E. Davis, and J. E. Davis, 1982. Ravens and buzzards in relation to sheep-farming and forestry in Wales. *J. Applied Ecol.* 19:681–706.
P. J. Dare, 1986. Raven *Corvus corax* populations in two upland regions of north Wales. *Bird Study* 33:179–89.
P. J. Ewins, J. N. Dymond, and M. Marquiss, 1986. The distribution, breeding and diet of ravens (*Corvus corax*) in Shetland. *Bird Study* 33:110–16.

153 *restricted to the immediate nest area:*
J. Gothe, 1961. Zur Ausbreitung und zum Fortpflanzungsverhalten des Kolkraben (*Corvus corax* L.) unter besonderer Berüksichtigung der Verhältnisse in Mecklenburg, in H. Schildmacher, *Beiträge zur Kenntnis deutscher Vögel.* Jena, Fischer, pp. 63–129.

153 *saw a pair escort a trio of strangers:*
J. Gothe, 1967. *Kolkrabe—schwarzer Gesell.* Hanover, Landbuch Verlag.

153 *mapped nearly all of the raven-nesting sites:*
J. Gothe, 1961. op. cit.

154 *the highest breeding density of ravens in the world:*
I. Newton, P. E. Davis, and J. E. Davis. 1982.

154 *in southern Scotland and northern England:*
M. Marquiss, I. Newton, and D. A. Ratcliffe, 1978. The decline
of the raven (*Corvus corax*) in relation to afforestation in southern Scotland and northern England. *J. Appl. Ecol.* 15:125–44.

154 *and in central Scotland:*
J. Mitchell, 1981. The decline of the raven as a breeding species
in central Scotland. *Forth Naturalist and Historian* 6:35–43.

154 *Hooper . . . and colleagues found:*
R. G. Hooper, H. S. Crawford, D. R. Chamberlain, and R. F.
Harlow, 1975. Nesting density of common ravens in the Ridge-
Valley region of Virginia. *Amer. Birds* 29:931–35.

154 *Shetland supports a dense population:*
P. J. Ewins, J. N. Dymond, and M. Marquiss, 1986. The distribu-
tion, breeding and diet of ravens *Corvus corax* in Shetland. *Bird
Study* 33:110–16.

155 *three pairs along one 1,000-yard stretch of cliff:*
D. T. Holyoak and D. A. Ratcliffe, 1968. The distribution of the
raven in Britain and Ireland. *Bird Study* 15:191–97.

155 *a 1.2 mile stretch seems to be the nearest distance:*
I. Newton, P. E. Davis, and J. E. Davis, 1982. op. cit.
P. J. Dare, 1986. op. cit.

155 *leave and never return:*
P. E. Davis and J. E. Davis, 1986. The breeding biology of a
raven population in central Wales. *Nature in Wales* 3:44–54.

155 *sleeping place by the nest:*
G. Kramer, 1932. Beobachtungen und Fragen zur Biologie des
Kolkraben (*Corvus c. corax* L.). *J. Ornith.* 80:329–42.
J. Gothe, 1961. op. cit.

155 *In northern Europe:*
J. Gothe, 1961. ibid.
J. L. Dorn, 1972. "The common raven in Jackson Hole, Wyo-
ming," M.S. Thesis. Univ. of Wyoming, Laramie, Wyo.

155 *observed even in the breeding season:*
D. K. Bryson, 1947. Large gathering of ravens during breeding
season. *Brit. Birds* 40:209, 41:19 (1948).
W. A. Cadman, 1947. A Welsh raven roost. *Brit. Birds* 40:209–10.
L. S. Young, K. A. Engel, A. Brody, and R. Bowman, 1985. Implications of communal roosting by common ravens to operation
and maintenance of the Malin to Midpoint 500 KV transmission
line. Snake River Birds of Prey Research Project Annual Report
1985, pp. 33–72. U.S. Dept of the Interior. Bureau of Land Management, Boise District, Idaho.

L. S. Young, K. A. Engel, and A. Brody, 1986. Implications of communal roosting by common ravens to operation and maintenance of the Malin to Midpoint 500 KV transmission line. Snake River Birds of Prey Research Project Annual Report 1986, pp. 36–59.

155 *breeding does not occur until:*
E. Gwinner, 1964. Untersuchungen über das Ausdrucks- und Sozialverhalten des Kolkraben (*Corvus corax* L.). *Z. Tierpsychol.* 21:657–748.

155 *swarms are the nonbreeding "surplus":*
D. T. Holyoak and D. A. Ratcliffe, 1968. op. cit.

155 *examples of "swarms" or flocks:*
J. Gothe, 1961. ibid.

156 *natural-history observations in Europe:*
J. Scheven, 1955. Ein Kolkrabenschwarm. *Vogelwelt* 76:212–16.
R. Hauri, 1956. Beiträge zur Biologie des Kolkraben (*Corvus corax*). *Ornith. Beob.* 53:28–35.
R. Hauri, 1958. Über Ansammlungen von Kolkraben (*Corvus corax*) im Berner Oberland. *Ornith. Beob.* 55:156–68.

156 *sightings . . . at dumps and other food sources:*
R. Melcher, 1949. Ist der Kolkrabe in den Schweizer Alpen häufiger geworden? *Ornithol. Beob.* 46:35–45.
J. Scheven, 1955. Ein Kolkrabenschwarm. *Vogelwelt* 76:212–16.
H. Meier, 1956. Kolkrabensammlungen im Mai. *Ornith. Beob.* 53:16.
G. Schmidt, 1957. Geselligkeit beim Kolkraben, insbesondere in Schleswig-Holstein. *Ornith. Mitt.* 9:121–26.
C. K. Mylne, 1961. Large flocks of ravens at food. *Brit. Birds* 54:206–207.

156 *a pioneer raven worker:*
R. Hauri, 1958. Über Ansammlungen von Kolkraben (*Corvus corax*) im Berner Oberland. *Ornith. Beob.* 55:156–68.

156 *berates Scheven:*
J. Scheven, 1955. Ein Kolkrabenschwarm. *Vogelwelt* 76:212–16.

156 *. . . and Schmidt:*
G. Schmidt, 1957. Geselligkeit beim Kolkraben, insbesondere in Schleswig-Holstein. *Ornith. Mitt.* 9:121–26.

157 *Clayton M. White and Merle Tanner-White from Brigham Young:*
C. M. White and M. Tanner-White, 1985. Unusual social feeding and soaring by the common raven (*Corvus corax*). *Great Basin Naturalist* 45:150–51.

157 *Gothe reports having seen several flocks:*
J. Gothe, 1961. op. cit.

348 NOTES

157 *Austin in his extensive survey:*
G. T. Austin, 1971. Roadside distribution of the common raven in the Mojave Desert. *Calif. Birds* 2:98.

157 *Hewson on several occasions saw:*
R. Hewson, 1949. Gathering of ravens. *Br. Birds* 42:181.

COMMUNAL ROOSTS

159 *birds at peak occupancy:*
J. Madson, 1976. The dance on Monkey Mountain. *Audubon* 78:52–60.

159 *Weffesten . . . admitted:*
J. Madson, 1976. ibid.

161 *Raven researchers Rolf Hauri . . . :*
R. Hauri, 1958. Über Ansammlungen von Kolkraben (*Corvus corax*) im Berner Oberland. *Ornith. Beob.* 55:156–68.

161 *. . . and W. A. Cadman from Wales:*
W. A. Cadman, 1947. A Welsh raven roost. *Brit. Birds* 40:209–10.

161 *Hutson observed a communal roost:*
H. O. W. Hutson, 1945. Roosting procedure of *Corvus corax laurenci* Hume. *Ibis* 87:455–59.

161 *as John Madson has put it:*
J. Madson, 1976. op. cit.

163 *March through July:*
L. S. Young, K. A. Engel, A. Brody, and R. Bowman, 1985. Implications of communal roosting by common ravens to operation and maintenance of the Malin to Midpoint 500 KV transmission line. Snake River Birds of Prey Research Project Annual Report 1985, pp. 33–72. U.S. Dept of the Interior. Bureau of Land Management, Boise District, Idaho.

163 *forage at night and gather into crowds:*
B. Heinrich and D. Vogt, 1980. Aggregation and foraging behavior of whirligig beetles (Coleoptera:Gyrinidae). *Behav. Ecol. & Sociobiol.* 7:179–86.

164 *the cultivated grain wastes on extensive fields:*
R. B. Stiehl, 1978. "Aspects of the ecology of the common raven in Harney Basin, Oregon." Ph.D. Thesis. Portland State Univ., Portland, Ore.
K. A. Engel, L. S. Young, W. G. Kell, and A. J. Brody, 1987. Implications of communal roosting by common ravens to operation and maintenance of the Malin to Midpoint 500 KV transmission line, in K. Steenhof, ed., Snake River Birds of Prey Report. U.S. Dept. of Interior, Bureau of Land Management, Boise, Idaho, pp. 34–55.

164 *reported on a roost of about two hundred birds:*
J. E. Cushing, Jr., 1941. Winter behavior of ravens at Tamales Bay, California. *Condor* 43:103–107.

164 *Ph.D. thesis work on ravens for Portland State:*
R. B. Stiehl, 1981. Observations of a large roost of common ravens. *Condor* 83:78.

164 *reports on a roost of ten ravens:*
S. A. Temple, 1974. Winter food habits of ravens on the arctic slope of Alaska. *Arctic* 27:41–46.

164 *report roosts of twenty-seven to seventy:*
C. K. Mylne, 1961. Large flocks of ravens at food. *Brit. Birds* 54: 206–07.

164 *found a roost of 106 ravens:*
V. J. Lucid and R. N. Conner. 1974. A communal common raven roost in Virginia. *Wilson Bull.* 86:82–83.

164 *tracked individual ravens:*
R. B. Stiehl. 1978. op. cit.

Do They Come from a Roost?

176 *soaring near a roost:*
R. B. Stiehl. 1978. op. cit.
V. J. Lucid and R. N. Conner. 1974. op. cit.

177 *European workers:*
R. Hauri, 1958. Über Ansammlungen von Kolkraben (*Corvus corax*) im Berner Oberland. *Ornith. Beob.* 55:156–68.
Beat Huber, 1988. Zur Sozialstruktur einer freilebenden Kolkrabenpopulation in der Schweiz. Master's Thesis, Zool. Institute Univ. Bern.

To Catch and Mark a Raven, or Two, or More

187 *continuous heat production:*
M. W. Schwan and D. D. Williams, 1978. Temperature regulation in the common raven of interior Alaska. *Comp. Biochem. Physiol.* 60:31–36.

Courting and Displays

198 *females in many diverse species:*
M. I. Avery, J. R. Krebs, and A. I. Houston, 1988. Economics of courtship-feeding in the European bee-eater (*Merops apiaster*). *Behav. Ecol. & Sociobiol.* 23:61–67.

350   NOTES

D. T. Gwynne, 1988. Courtship feeding in katydids benefits the
mating male's offspring. *Behav. Ecol. & Sociobiol.* 23:373–77.

199 *illustrated in Franklin Coombs's* . . . :
F. Coombs, 1978. *The Crows: A Study of the Corvids of Europe.*
London, B. T. Batsford.

199 *and Derek Goodwin's books* . . . :
D. Goodwin, 1986. *Crows of the World,* 2nd ed. Seattle, Univ. of
Wash. Press.

199 *the German edition:*
K. Lorenz, 1940. Die Paarbildung beim Kolkraben. *Z. Tierpsychol.* 3:278–92.

199 *made from these photographs:*
K. Lorenz. 1968. Pair formation in ravens, ed., Heinz Friedrich,
Man and Animals: Studies in Behavior. New York, St. Martin's
Press, pp. 17–36.
J. Gothe, 1967. *Kolkrabe-schwarzer Gesell.* Hanover, Landbuch
Verlag.

199 *the classic work:*
E. Gwinner, 1964. Untersuchungen über das Ausdrucks- und
Sozialverhalten des Kolkraben (*Corvus corax* L.). *Z. Tierpsychol.*
21:657–748.
J. Gothe. 1961. Zur Ausbreitung und zum Fortpflanzungsver-
halten des Kolkraben (*Corvus corax* L.) unter besonderer Berück-
sichtigung der Verhältnisse in Mecklenburg, in H. Schildmacher,
Jena, Fischer, *Beiträge zur Kenntnis deutscher Vogel.* pp. 63–129.

200 *well-known scientists who have studied ravens:*
O. and M. Heinroth, 1926. *Die Vögel Mitteleuropas,* vol. 1. Ber-
lin, H. Bermühler.
G. Kramer, 1932. Beobachtungen und Fragen zur Biologie des
Kolkraben (*Corvus c. corax* L.). *J. Ornith.* 80:329–42.
K. Lorenz, 1968. op. cit.
E. Gwinner, 1964. op. cit.
J. Gothe, 1963. Zur Droh- und Beschwichtigungsgebärde des
Kolkraben (*Corvus c. corax* L.) *Z. Tierpsychol.* 19:687–91.

202 *copulatory solicitation:*
E. Gwinner, 1964. op. cit.

203 *made systematic observations:*
D. Van Vuren, 1984. Aerobic rolls by ravens on Santa Cruz Island, California. *Auk* 101:620–21.

203 *pass objects back and forth:*
R. N. Conner, D. R. Chamberlain, and V. J. Lucid, 1973. Some
aerial flight maneuvers of the common raven in Virginia. *The
Raven.* 44:99.

203 *describes observations near his cabin:*
F. Zirrer. 1945. The Raven. *Passenger Pigeon.* 7:61–67.

203 *four different kinds of flight:*
J. Gothe, 1961. op. cit.

204 *as described by Emeis:*
W. Emeis, 1951. Beobachtungen im Brutgebiet des Kolkraben. *Orn. Mitt.* 3:217–70, 241–46.

204 *and Van Vuren:*
D. Van Vuren, 1984. Ibid.

204 *whereas Van Vuren:*
D. Van Vuren, 1981. op. cit.

204 *Gothe suggests . . . :*
J. Gothe, 1963. op. cit.

204 *Van Vuren noted:*
D. Van Vuren, 1981. op. cit.

204 *few other ravens:*
J. Gothe, 1961. op. cit.
D. Van Vuren, 1984. op. cit.

204 *crowds of forty or more:*
D. K. Bryson, 1947. Large gathering of ravens during breeding season. *Brit. Birds* 40:209, 41:19 (1948).
W. A. Cadman, 1947. A Welsh raven roost. *Brit. Birds* 40:209–210.
J. J. B. Young, 1949. Flocking of ravens. *Brit. Birds* 42:151.
A. G. Hurrell, 1951. Ravens using thermals. *Brit. Birds* 44:88–89.

205 *describes numerous variations:*
E. Gwinner, 1966. Uber einige Bewegungsspiele des Kolkrabens (*Corvus c. corax L.*) *Z. Tierpsychol.* 23:28–36.
R. D. Elliot, 1977. Hanging behavior in common ravens. *Auk* 94:777–778.

205 *hanging ravens . . . attacked:*
E. Gwinner, 1966. ibid.

206 *useful for rearing young:*
J. T. Lifjeld and T. Slagsvold, 1986. The function of courtship feeding during incubation in the pied flycatcher *Ficedula hypoleuca. Anim. Behav.* 34:1441–53.

206 *"rrock" calls:*
D. Van Vuren, 1984. op. cit.

INDIVIDUALS

220 *herring gulls:*
J. A. Kodlec and W. H. Drury, 1968. Structure of the New England herring gull population. *Ecology* 49:644–76.

220 *and pinyon jays:*
J. M. Marzluff and R. P. Balda, 1987. Resource and climatic variability: influence on sociality of two southwestern corvids, C. N. Slobodchikoff, ed., *The Ecology of Social Behavior.* New York, Academic Press, pp. 255–83.

RAVEN CALLS

246 *describes seeing two ravens in Alaska:*
F. Bruemmer, 1984. Ravens. *International Wildlife* 14:33–35.

246 *ornithologist Thomas Nuttall remarked:*
T. Nuttall, 1903. *Birds of the United States and Canada.* Boston, Little, Brown & Co.

247 *Eleanor D. Brown discovered:*
E. D. Brown, 1985. The role of song and vocal imitation among common crows (*Corvus brachyrhynchos*). Z. *Tierpsychol.* 68:115–136.

247 *his publication provides sonograms:*
E. Gwinner, 1964. Untersuchungen uber das Ausdrucks- und Sozialverhalten des Kolkraben (*Corvus corax* L.) Z. *Tierpsychol.* 21:657–748.

247 *1972 thesis works of Jane L. Dorn:*
J. L. Dorn, 1972. "The common raven in Jackson Hole, Wyoming," M.S. Thesis. Laramie, Univ. of Wyoming.

247 *and Roderick N. Brown:*
R. N. Brown, 1974. "Aspects of vocal behavior of the raven (*Corvus corax*) in interior Alaska." M.Sc. Thesis. Fairbanks, University of Alaska.

248 *Connor from Virginia Polytechnic Institute:*
R. Conner, 1985. Vocalizations of common ravens in Virginia. *Condor* 87:379–388.

248 *eventually published in sonograms:*
B. Heinrich, 1988. Winter foraging at carcasses by three sympatric corvids, with emphasis on recruitment by the raven, *Corvus corax. Behav. Ecol. & Sociobiol.* 23:141–156.

248 *Zirrer described the distinctive call:*
F. Zirrer, 1945. The Raven. *Passenger Pigeon* 7:61–67.

248 *Roderick Brown heard:*
R. N. Brown, 1974. op. cit.

248 *which Gwinner considers:*
E. Gwinner, 1964. op. cit.

249 *corvid specialist Ian Rowley:*
I. Rowley, 1973. The comparative ecology of Australian corvids. *CSIRO Wildl.* 18:25–65.

249 *Brown observed it:*
R. N. Brown, 1974. op. cit.

249 *Andrieux has provided a sonogram:*
M. Andrieux, 1984. La prosodie dans la semantique du cri d'alarme du grand corbeau (*Corvus corax* L.) *Biol. of Behav.* 9:59–63.

249 *Angell also describes "sentry" ravens:*
T. Angell, 1978. *Ravens, Crows, Magpies, and Jays.* Seattle, Univ. of Washington Press.

249 *Gwinner's separate groups:*
E. Gwinner, 1964. op. cit.

249 *individual-specific call:*
E. Gwinner and J. Kneutzen, 1962. Uber die biologische Bedeutung der "zweckdienlichen" Anwendung erlernter Laute bei Vogeln. *Z. Tierpsychol.* 19:692–696.

249 *Gwinner's raven Wotan:*
E. Gwinner, 1964. ibid.

249 *soft "korr" used at the nest:*
J. Gothe, 1961. Zur Ausbreitung und zum Fortpflanzungsverhalten des Kolkraben (*Corvus corax* L.) unter besonderer Berücksichtigung der Verhältnisse in Mecklenburg, in H. Schildmacher, *Beiträge zur Kenntnis deutscher Vögel.* Jena, Fischer, pp. 63–129.

250 *both Konrad Lorenz:*
K. Lorenz, 1952. *King Solomon's Ring: New Light on Animal Ways.* New York, Crowell.

250 *and Tony Angell:*
T. Angell, 1978. ibid.

250 *Nelson quotes Koyukon native hunters:*
R. K. Nelson, 1983. *Make Prayers to the Raven: A Koyukon view of the Northern Forest.* Chicago, Univ. of Chicago Press.

252 *Lorenz, citing no evidence:*
K. Lorenz, 1952. ibid.

252 *more than any other bird I know:*
A. C. Bent, 1964. *Life Histories of North American Jays, Crows, and Titmice.* Part I. New York, Dover.

252 *Zirrer calls the song:*
F. Zirrer, 1945. op. cit.

Why Be Brave?

268 *"Wolves seem to pay":*
R. K. Nelson, 1983. *Make Prayers to the Raven: A Koyukon view of the Northern Forest.* Chicago, Univ. of Chicago Press.

269 *saw ravens give way:*
J. L. Dorn, 1972. "The common raven in Jackson Hole, Wyoming," M.S. Thesis. Univ. of Wyoming, Laramie, Wyo.

269 *Hawkins describes an episode:*
E. Hawkins, 1988. The wolf at the window. *Audubon* 90:116.

269 *wildlife ecologist R. O. Pedersen:*
R. O. Pedersen, 1977. Wolf ecology and prey relationships on Isle Royale. Nat. Park Serv. Sci. Mon. Ser. No. 11.

269 *The Heinroths observed:*
O. and M. Heinroth, 1926. *Die Vögel Mitteleuropas,* vol. 1. Berlin, H. Bermühler.

270 *saw six ravens land:*
D. Bruggers, 1985. Raven (Mis)Behavior. Bell Museum of Nat. History Imprint II:1–8.

271 *the male courting raven:*
E. Gwinner, 1964. Untersuchungen über das Ausdrucks- und Sozialverhalten des Kolkraben (*Corvus corax corax* L.). *Z. Tierpsychol.* 21:657–748.
K. Lorenz, 1968. Pair formation in ravens, in Heinz Friedrich, ed., *Man and Animals: Studies in Behavior.* New York, St. Martin's Press, pp. 17–36.

271 *a study by Tore Slagsvold:*
T. Slagsvold, 1984. The mobbing behaviour of the hooded crow *Corvus corone cornix:* anti-predator defence or self-advertisement? Fauna norv. ser. c. circlus 7:127–31.
T. Slagsvold, 1985. Mobbing behaviour of the hooded Crow *Corvus corone cornix* in relation to age, sex, size, season, temperature and kind of enemy. *Fauna norv. Ser. C. Circlus* 8:9–17.

271 *L. David Mech specifically cites:*
L. D. Mech, 1970, pp. 279–287. *The Wolf: The Ecology and Behavior of an Endangered Species.* Garden City, N.Y., Natural History Press.

Tradeoffs and Complexities

280 *Gwinner describes how:*
E. Gwinner, 1964. Untersuchungen über das Ausdrucks- und Sozialverhalten des Kolbraben (*Corvus corax* L.). *Z. Tierpsychol.* 21:657–748.

INDEX

(Page numbers in *italics* refer to data compiled in graphs.)

acrobatic flight, 186, 218
  in courting, 203–5, 206–7
  as play behavior, 204, 206–7
Adamant, Vt., nest near,
  232–38, 242–45
Adams, Billy, 290
adult pairs:
  baits defended by, 151–52,
    223–24, 230, 254–60, 264–
    266, 277, 278, 279, 281, 311
  behavior of, 210, 297
  feather postures of, 210,
    310–11
  juveniles chased from bait
    by, 211, 212–13, 214, 215–
    216, 223–24, 230–31, 255–
    257, 277, 278, 310, 311
  permanent bonds of, 153
  quorks as territorial adver-
    tisement calls of, 224–25,
    230, 231, 249, 254–56, 257
    259, 264, 265, 277, 279,
    281, 283, 284, 285, 289, *318*
  tender exchanges of, 212,
    237, 244, 250
  territory or domain of, 151–
    152, 153–55, 165, 186
adults:
  communal roosts joined by,
    165
  fear of baits in juveniles vs.,
    215–16, 310
  feathers of juveniles vs., 126

juveniles' fear of, 223–24
juveniles' submissiveness
  with, 217
mouth color of, 144
size inadequate as distinguishing characteristic of,
  333
unmated, feather postures of,
  213
unmated, low status of,
  217–18
aerial acrobatics, *see* acrobatic
  flight
aerial chases, *see* chases, aerial
African brown-necked ravens
  (*Corvus ruficollis*), 17
African honeyguides, 63
age distribution, 220, 295
aggressiveness, 121–22
  in aerial chases, 170, 171
  among crows, 122
  recruitment and, 225–26
  see also confrontations;
    fighting
Ahab, King, 23
Aleutian Islands, 21
Allen, Durward L., 55–56
altruism:
  reciprocal, 105, 110, 213,
    308, 309
  see also food sharing;
    sharing behavior
Amadon, Dean, 18

crows, *continued*
  communication between
    ravens and, 174
  foraging behavior of, 66,
    166–67, 169–70, 173, 227,
    263, 296, 302–3, *317, 329,*
    331
  as frugivores, 77
  illegality of tampering with
    nests of, 28–29
  during incubation period,
    227
  intelligence of, 113, 114–15
  migration of, 86
  New Caledonia (*Corvus
    manaduloides*), 114
  Northwestern (*Corvus
    caurinus*), 114
  pairing of, 86
  persecution of, 27–29
  ravens' interactions with,
    174, 225, 227
  recruitment in ravens vs.,
    76–77
  used as sentinels, 169, 173,
    174
  vocalizations of, 169, 173,
    174, 227, 263
Crows of the World
  (Goodwin), 248
Curious Naturalists
  (Tinbergen), 51, 56
Cushing, J. E., Jr., 164
Cyanocitta cristata (common
  blue jays), 114

danger, *see* risk
Danish crows (*Corvus corone*),
  17, 153, 160, 161, 271
Davida (raven), 200
Davis, J. E., 154
Davis, P. E., 154
death:
  feigned by carnivores, 76
  ravens associated with, 22
Death of the Birdman Scene
  (Lascaux), 21

decoys, stuffed ravens as, 98,
  100, 126, 227–28
deer, 53, 62, 84, 308
Denali, Mount (Mount
  McKinley), 49–50
density:
  of crowds, fighting related
    to, 122, 171, 174, 315, *334*
  of raven-nesting sites,
    153–55
desert crows (*Corvus rufi-
    collis*), 17
desert tortoises, 50–51
Dickkopf ("thickhead"
  display), 200, 201–2
DiDomenico, Litia, 191
DiSotto, Henry, 289, 295
DiSotto, Leona, 289
displays, 158, 199–207
  acrobatic flight and, 203–5,
    206–7
  bowing motions in, 200–201,
    202, 209, 212, 270–71,
    296–97
  courtship, 148–49, 175, 199,
    201, 203–7, 209–10, 270–
    271, 296–97
  of dominance, 148–49, 249,
    270
  leading hunters to prey,
    251–52
  self-assertive (*Imponier-
    verhalten*), 200–202, 247,
    297
  of submission, 202, 229
  "thickhead" (*Dickkopf*),
    200, 201–2
dives, 203, 204, 218
Dixfielders (ravens), 139, 140,
  141
DNA, 19
dogs, 84
  pestered by ravens, 271
  sled, 268
dominance:
  breeding privileges and, 145,
    312
  displays of, 148–49, 249, 270

ABOUT THE AUTHOR

Bernd Heinrich is Professor Emeritus of Biology at the University of Vermont and is the author of numerous books, including *Bumblebee Economics, Why We Run, Mind of the Raven,* and most recently, *The Homing Instinct.* His work has been featured in *Scientific American, Discover,* and *The New York Times,* among other places. He has received the John Burroughs Medal for nature writing and has been nominated for a National Book Award for Science. He lives in Vermont.